国际工程科技发展战略高端论坛

International Top-level Forum on Engineering Science and Technology Development Strategy

中国工程院
CHINESE ACADEMY OF ENGINEERING

# 未来的化工冶金材料工程

*WEILAI DE HUAGONG YEJIN CAILIAO GONGCHENG*

# THE FUTURE OF CHEMICAL, METALLURGY AND MATERIAL ENGINEERING

高等教育出版社·北京

内容提要

2014年6月2~3日，2014年国际工程科技大会在北京召开。大会由联合国教育、科学及文化组织（UNESCO），国际工程与技术科学院理事会（CAETS）和中国工程院联合举办。“未来的化工、冶金与材料工程”分会由中国工程院化工、冶金与材料工程学部和北京化工大学承办，来自国内外化工、冶金及材料领域共60余位中外院士，80余位相关学者、行业专家共计150人参加了该分会。书中收录了该分会的相关报告及摘要，内容涵盖了先进材料、生物技术和计算机模拟体系、冶金技术等各研究方向的最新进展及成果，探讨了化工、冶金与材料工程领域的研究现状及面临的问题与挑战，有助于推动相关领域的研究与合作。

本书中的论述和分析，对化工、冶金、材料相关领域的科研人员、技术人员以及研究生具有重要的参考价值。

**图书在版编目(CIP)数据**

未来的化工冶金材料工程：汉英对照/中国工程院编著. - 北京：高等教育出版社，2016.2
(国际工程科技发展战略高端论坛)
ISBN 978-7-04-044023-2

Ⅰ. ①未… Ⅱ. ①中… Ⅲ. ①化工材料-材料科学-研究-汉、英②冶金-材料-材料科学-研究-汉、英 Ⅳ. ①TQ04②TF03

中国版本图书馆CIP数据核字(2016)第245229号

**总 策 划** 樊代明
策划编辑 王国祥 黄慧靖　　责任编辑 沈晓晶
封面设计 顾 斌　　责任印制 韩 刚

出版发行 高等教育出版社
社 址 北京市西城区德外大街4号
邮政编码 100120
印 刷 北京汇林印务有限公司
开 本 850mm×1168mm 1/16
印 张 10.25
字 数 268千字
购书热线 010-58581118
咨询电话 400-810-0598
网 址 http://www.hep.edu.cn
http://www.hep.com.cn
网上订购 http://www.landraco.com
http://www.landraco.com.cn
版 次 2016年2月第1版
印 次 2016年2月第1次印刷
定 价 80.00元


物 料 号 44023-00

# 目　　录

## 第一部分　综述



## 第二部分　参会专家名单



## 第三部分　主题报告及报告人简介

# CONTENTS

## Part Ⅰ Overview of the Top-level Forum



## Part Ⅱ List of Experts Attending the Forum



## Part Ⅲ Keynote Speech and Speaker Introduction

# 第一部分
# 综　　述

# 综　　述

2014年6月2~3日,2014年国际工程科技大会在北京召开。本次大会由联合国教育、科学及文化组织(UNESCO),国际工程与技术科学院理事会(CAETS)和中国工程院联合举办。大会主题为“工程科技与人类未来”,意在为来自全球工程界、产业界、研究机构以及政府的与会者提供一个论坛,分享工程科技前沿新知,探索未来发展方向,为应对未来人类所面临的挑战聚集智慧。来自全球30多个国家的国家工程院院长、中国工程院院士和外籍院士、中外工程科技界代表等1500多人参加了这一工程科技界的盛会,此次会议共分“未来机械工程”、“信息网络与社会发展”、“未来的化工、冶金与材料工程”等九个分会。

由中国工程院化工、冶金与材料工程学部和北京化工大学承办的“未来的化工、冶金与材料工程”分会于6月2日、3日下午在北京会议中心召开,来自国内外化工、冶金及材料领域60余位中外院士,80余位相关学者、行业专家共计150人参加了该分会。屠海令院士在分会开幕式致辞中指出,工程科技对推动经济发展、社会进步和提高人民生活水平都做出了重要贡献,此次会议主要研讨工程科技的创新与人类未来如何为全球经济和社会的可持续发展做出新贡献,并希望通过此次会议帮助我们进一步加强全球范围内科学界、工程界之间的合作。

专家报告会分别由谭天伟、Ulrich W. Suter、刘炯天、余艾冰、徐惠彬、徐政和、周玉和Robin J. Batterham教授主持。中国科学院薛其坤院士做了题为“量子材料生长的原子水平控制:从量子反常霍尔效应到高温超导”的报告。美国国家工程院 Cato T. Laurencin 院士和澳大利亚科学院余艾冰院士,分别做了题为“再生工程,一个全新的领域:理论与实践”和“粒子系统的模拟与建模”的学术报告。来自美国国家工程院的范良士院士(化学循环技术:俄亥俄州立大学铁基过程),中国工程院的付贤智院士(基于光催化的清洁能源与环境新技术)、张兴栋院士(可诱导组织再生的生物材料——生物材料发展的新纪元)和邱冠周院士(用生物技术的钥匙开启矿产资源利用的大门),英国皇家工程院的 Helen Valerie Atkinson 院士(玉米淀粉、番茄酱与汽车零配件:半固态加工技术综述),印度国家工程院的 K. V. Raghavan 院士(化学与材料工程领域未来产品与工艺研发的挑战),加拿大工程院的徐政和院士(能源与矿产资源的发展和利用:过去、现在及未来),瑞士工程科学院的 Arthur Ruf 院士(流程优化的系统方法)分别从各自的研究领域阐述了化工、冶金、材料工程的最新进展,研讨了工程科技的创新与人类未来,并就如何为全球经济和社会的可持续发展做出新贡献进行了讨论。学术报告结束后,与会专家围绕报告中介绍的先进技术及工程案例进行了广泛深入的交流,并对与工程科技相关的教育、人才培养、科技创新等主题交换了意见。

屠海令院士做分会总结发言,他指出“未来的化工、冶金与材料工程”分会学术报告涵盖了先进材料、生物技术和计算机模拟体系、冶金技术等各研究方向的最新进展及成果,并且通过大家积极热烈的讨论,使我们更好地了解化工、冶金与材料工程领域的研究现状及面临的问题与挑战,这必将有助于推动相关领域的研究与合作。

# 第二部分

# 参会专家名单

# 参会专家名单

**Ulrich W. Suter** 瑞士工程科学院,瑞士工程科学院院士

**范良士** 俄亥俄州立大学,美国国家工程院院士,中国工程院外籍院士

**Arthur Ruf** 瑞士苏黎世联邦高等工学院,瑞士工程科学院院士

**Cato T. Laurencin** 康涅狄格大学,美国国家工程院院士

**Helen Valerie Atkinson** 莱斯特大学,英国皇家工程院院士

**K. V. Raghavan** 印度化学技术研究所,印度工程院院士

**徐政和** 艾伯塔大学,加拿大工程院院士

**余艾冰** 莫纳什大学,澳大利亚科学院和技术科学与工程院院士

**Robin J. Batterham** 力拓公司,澳大利亚工程院院士,中国工程院外籍院士

**Eric Forssberg** 吕勒奥理工大学,瑞典工程科学院院士,中国工程院外籍院士

**P. Somasundaran** 哥伦比亚大学,美国国家工程院院士,中国工程院外籍院士

**Hannelore Bowman** 化学工程师学会,教授

**Yang Shen** 哥伦比亚大学,教授

**薛其坤** 清华大学,中国科学院院士

**曹湘洪** 中国石油化工集团公司,中国工程院院士

**陈祥宝** 中国航空工业集团公司北京航空材料研究院,中国工程院院士

**戴永年** 昆明理工大学,中国工程院院士

**付贤智** 福州大学,中国工程院院士

**傅恒志** 西北工业大学,中国工程院院士

**干　勇** 中国工程院/钢铁研究总院,中国工程院院士

**高从堦** 杭州水处理技术研究开发中心有限公司,中国工程院院士

**何季麟** 中色(宁夏)东方集团公司,中国工程院院士

**胡永康** 中国石油化工股份有限公司抚顺石油化工研究院,中国工程院院士

**黄伯云** 中南大学,中国工程院院士

**蹇锡高** 大连理工大学,中国工程院院士

**江东亮** 中国科学院上海硅酸盐研究所,中国工程院院士

**柯　伟**　中国科学院金属研究所,中国工程院院士
**李冠兴**　中核北方核燃料元件有限公司,中国工程院院士
**李龙土**　清华大学,中国工程院院士
**李言荣**　电子科技大学,中国工程院院士
**李元元**　吉林大学,中国工程院院士
**李仲平**　航天材料及工艺研究所,中国工程院院士
**刘伯里**　北京师范大学化学学院,中国工程院院士
**刘炯天**　郑州大学,中国工程院院士
**毛炳权**　中国石油化工股份有限公司北京化工研究院,中国工程院院士
**钱旭红**　华东理工大学,中国工程院院士
**邱冠周**　中南大学,中国工程院院士
**桑凤亭**　中国科学院大连化学物理研究所,中国工程院院士
**舒兴田**　中国石油化工集团公司石油化工科学研究院,中国工程院院士
**孙传尧**　北京矿冶研究总院,中国工程院院士
**谭天伟**　北京化工大学,中国工程院院士
**屠海令**　煤炭科工集团有限公司,中国工程院院士
**汪燮卿**　中国石油化工股份有限公司石油化工科学研究院,中国工程院院士
**汪旭光**　北京矿冶研究总院,中国工程院院士
**王淀佐**　中国工程院/北京有色金属研究总院,中国工程院院士
**王海舟**　钢铁研究总院,中国工程院院士
**王静康**　天津大学化工学院,中国工程院院士
**王一德**　太原钢铁(集团)有限公司,中国工程院院士
**翁宇庆**　中国金属学会,中国工程院院士
**吴慰祖**　总参谋部第五十五研究所,中国工程院院士
**吴以成**　中国科学院理化技术研究所,中国工程院院士
**徐承恩**　中国石化工程建设公司,中国工程院院士
**徐德龙**　西安建筑科技大学,中国工程院院士
**徐惠彬**　北京航空航天大学,中国工程院院士
**徐匡迪**　全国政协/中国工程院,中国工程院院士
**杨启业**　中国石化工程建设公司,中国工程院院士
**殷国茂**　成都无缝钢管有限责任公司,中国工程院院士
**殷瑞钰**　钢铁研究总院,中国工程院院士
**袁晴棠**　中国石油化工集团公司,中国工程院院士

**袁渭康** 华东理工大学,中国工程院院士
**张国成** 北京有色金属研究总院,中国工程院院士
**张寿荣** 武汉钢铁(集团)公司,中国工程院院士
**张文海** 中国瑞林工程技术有限公司,中国工程院院士
**张兴栋** 四川大学国家生物医学材料工程技术研究中心,中国工程院院士
**赵连城** 哈尔滨工业大学光电信息科学系,中国工程院院士
**周　廉** 西北有色金属研究院,中国工程院院士
**周　玉** 哈尔滨工业大学,中国工程院院士
**左铁镛** 北京工业大学,中国工程院院士
**曲　涛** 昆明理工大学,教授
**杨　骥** 中国工程院,教授
**刘祖铭** 中南大学,教授
**姚国成** 中国工程院,教授
**赵　雷** 钢铁研究总院,教授
**龚俊波** 天津大学化工学院,教授
**段国瑞** 太原钢铁(集团)有限公司,教授
**杨嘉伟** 西安建筑科技大学,教授
**郭继东** 中国工程院,教授
**张旭孝** 钢铁研究总院,教授
**张文皓** 武汉钢铁(集团)公司,教授
**詹小青** 中国瑞林工程技术有限公司,教授
**张　璇** 四川大学国家生物医学材料工程技术研究中心,教授
**朱宏康** 西北有色金属研究院,教授
**吴玉峰** 北京工业大学,教授
**赵学良** 中国石油化工集团公司,教授
**陈　鹰** 北京师范大学,教授
**秦培勇** 北京化工大学,教授
**陈必强** 北京化工大学,教授
**张　栩** 北京化工大学,教授
**刘　静** 中国科学院理化技术研究所,教授
**林哲帅** 中国科学院理化技术研究所,教授
**高宏伟** 中国科学院理化技术研究所,教授
**龙　军** 石油化工科学研究院,教授

郭锦标　石油化工科学研究院,教授
傅　军　石油化工科学研究院,教授
田松柏　石油化工科学研究院,教授
吴　巍　石油化工科学研究院,教授
慕旭宏　石油化工科学研究院,教授
代振宇　石油化工科学研究院,教授
侯栓弟　石油化工科学研究院,教授
崔龙鹏　石油化工科学研究院,教授
程　薇　石油化工科学研究院,教授
郭湘波　石油化工科学研究院,教授
张宝吉　石油化工科学研究院,教授
桂夏辉　中国矿业大学,教授
闫小康　中国矿业大学,教授
范桂侠　中国矿业大学,教授
万克记　中国矿业大学,教授
李国胜　中国矿业大学,教授
王　爱　中国矿业大学,教授
邢耀文　中国矿业大学,教授
孔小燕　中国矿业大学,教授
苗真勇　中国矿业大学,教授
张志军　中国矿业大学(北京),教授
黄　根　中国矿业大学(北京),教授
韩桂洪　郑州大学,教授
黄艳芳　郑州大学,教授
黄松涛　北京有色金属研究总院,副院长
卢世刚　北京有色金属研究总院,副总工程师
朱　强　北京有色金属研究总院,副总工程师
王立根　北京有色金属研究总院,副总工程师
常秀敏　北京有色金属研究总院,主任
温建康　北京有色金属研究总院,主任
米绪军　北京有色金属研究总院,主任
蒋利军　北京有色金属研究总院,主任
车小奎　北京有色金属研究总院,主任

宋永胜　北京有色金属研究总院,副主任
杨志民　北京有色金属研究总院,副所长
李腾飞　北京有色金属研究总院,副主任
张　倩　钢铁研究总院,教授
崔怀周　钢铁研究总院,教授
刘质斌　钢铁研究总院,教授
童金涛　钢铁研究总院,教授
唐　超　钢铁研究总院,教授
冯　硕　钢铁研究总院,教授
沈雯雯　钢铁研究总院,教授
张子阳　钢铁研究总院,教授
谭清元　钢铁研究总院,教授
徐小青　钢铁研究总院,教授
张正延　钢铁研究总院,教授
史晓强　钢铁研究总院,教授
马亚鑫　钢铁研究总院,教授
刘　雨　钢铁研究总院,教授
许林皓　钢铁研究总院,教授
康　峰　钢铁研究总院,教授
范新超　钢铁研究总院,教授

# 第三部分

# 主题报告及报告人简介

# 量子材料生长的原子水平控制：从量子反常霍尔效应到高温超导


薛其坤

清华大学，北京



**摘要**：分子束外延（MBE）是一个用于制备半导体及相关异质结构的功能强大的技术，高迁移率的二维电子气和多量子阱结构级联激光器是其中两个著名的例子。用扫描隧道显微镜（STM）和角分辨光电子能谱（ARPES）结合 MBE 可以使它的能量到达到前所未有的程度。在这次演讲中，我将展示我们在有磁性掺杂的拓扑绝缘体情况下的反常霍尔效应和用 MBE-STM-ARPES 解决高温超导电性的途径的问题。

**薛其坤**　1963 年生于山东，1984 年毕业于山东大学光学系激光专业，1994 年在中国科学院物理研究所获得博士学位。1992—1999 年先后在日本东北大学金属材料研究所和美国北卡莱罗纳州立大学物理系学习和工作。1999—2007 年任中国科学院物理研究所研究员、课题组组长，1999—2005 年任表面物理国家重点实验室主任。2005 年起任清华大学物理系教授，同年 11 月被增选为中国科学院院士。2010—2013 年任清华大学理学院院长、物理系主任，2011 年起任低维量子物理国家重点实验室主任，2013 年 5 月起任清华大学分管科研的副校长。是国际著名期刊 *Surface Science Reports*、*Physics Review B*、*Applied Physics Letters*、*Journal of Applied Physics* 和 *AIP Advances* 等的编委，*Nano Research* 和 *Surface Review & Letters* 的主编。薛其坤

是国际著名的实验物理学家，其主要研究方向为扫描隧道显微学、表面物理、自旋电子学、拓扑绝缘量子态、低维超导电性等。发表文章360余篇，包括5篇*Science*，9篇*Nature*子刊，31篇*Physical Review Letters*，被引用超过7400余次。在国际会议上应邀做大会/主题/特邀报告100余次，其中，五次在美国物理学会年会做邀请报告。曾获何梁何利科学与进步奖(2006年)、国家自然科学奖二等奖(2005年、2011年)、第三世界科学院物理奖(2010年)、求是杰出科技成就集体奖(2011年)、陈嘉庚科学奖(2012年)和"万人计划"杰出人才(2013年)等奖励与荣誉。

# 再生工程，一个全新的领域：理论与实践

**Cato T. Laurencin**

康涅狄格大学，斯托斯，康涅狄格，美国

**摘要**：未来十年我们将看到再生肌肉骨骼组织会有前所未有的进步。我们正从先进的修复时代，到我称之为再生工程的时代。在此过程中，我们有能力解决肌肉骨骼再生的重大挑战。例如，骨、韧带、软骨等组织，我们对它们的了解从细胞水平到组织水平。通过组织工程技术，我们有能力临床生产这些组织。并且已经在优化工程组织方面取得了进步，部分原因是由于先进的材料科学/纳米技术和干细胞技术这两个比较新的工程领域的进步。关键参数影响组织再生的新模型设计。除了我们对材料了解水平的进步，了解细胞和完整的组织行为及如何通过材料的设计进行调制也将是很重要的。必须用一个完整而全面的方法来替换再生组织的设计系统，同时细胞、生物因素、支架和形成胚胎等因素也相当重要。

**Cato T. Laurencin** 美国工程院院士，美国科学院医学研究所成员。获得普林斯顿大学化学工程专业的学士学位，哈佛医学院的硕士学位，麻省理工学院生物化学工程和生物技术的博士学位。Laurencin 教授是康涅狄格大学教授，康涅狄格大学临床与转化科学研究所的首席执行官。他是学校著名的骨科手术教授、化学和分子生物学教授、材料工程教授、生物医学教授。他同时是康涅狄格大学再生工程研究所及雷蒙德和贝弗利萨克中心的创始人和主任。他是再生工程领域的开拓者，研究领域包括先进生物材料科学、纳米技术、干细胞科学和组织再生技术。他是美国材料

研究学会会员、国际生物材料科学与工程学会会员、生物医学工程会员，并且被美国化学工程师学会在100周年庆典时称为“现代100名工程师之一”。Laurencin教授被美国克林顿总统授予“总统研究员奖”，被奥巴马总统授予“总统杰出导师奖”。

# 粒子系统的模拟与建模

余艾冰

莫纳什大学,墨尔本,维多利亚,澳大利亚

**摘要**:粒子科学技术是一个迅速发展的跨学科研究领域,其核心是对颗粒/颗粒物质在微观和宏观的性质的理解,这些是我们可以看到的物质的状态,但是我们对它并不了解。

颗粒物质表现出的宏观行为,是由单个粒子之间的相互作用,和粒子与周围的气体、液体及器壁的相互作用共同控制的。这一目标可以通过颗粒的规模研究而实现,而颗粒规模研究基于具体的微观动态信息,比如对个体颗粒作用力和运动轨迹的详细研究。近年来,这样的研究在全球得到了迅速的发展,这主要是由于独立粒子模拟技术及计算机技术的飞速发展。本报告将介绍我在这个方向的工作概况,涵盖了不同条件下的理论发展和案例研究。通过代表性的例子演示了在化工、冶金与材料工程中,已经证明小颗粒的研究与科学中许多具有挑战性的问题有很密切的联系。不仅是基础研究而且在工程应用中,颗粒规模研究已经逐渐变成一个强大的工具。最后,我简要讨论一下未来发展的领域。

**余艾冰** 澳大利亚科学院和技术科学与工程院院士,澳大利亚新南威尔士大学教授,东北大学长江学者特聘教授。分别于 1982 年和 1985 年获得东北大学冶金工艺学专业的学士和博士学位;1990 年获得澳大利亚卧龙岗大学的博士学位;2007 年获澳大利亚新南威尔士大学 DSc 学位。现为澳大利亚新南威尔士大学教授,领导一个世界一流的研究设施“颗粒体系计算机仿真与模拟”(SIMPAS)。担任澳大利亚-中国矿物

冶金和材料研究中心主任。2014 年 4 月担任澳大利亚莫纳什大学副校长、兼 Monash-东南大学联合研究院院长。余教授是颗粒/粉末技术和过程工程领域的国际著名科学家,已经发表论文 750 余篇,其中,ISI 收录 430 余篇,受邀参加多次国际会议并作主题报告。2008—2013 年,担任 *Particuology* 期刊主编;2013 年,担任 *Powder Technology* 期刊主编;2014 担任 *Powder Science and Engineering* 期刊主编,并且一直担任 20 多个学术期刊的编委。曾荣获"澳大利亚研究委员会伊丽莎白女王二世奖学金""澳大利亚专业和联邦奖学金",荣获澳大利亚科学院钢铁协会"伊恩 · 华克奖章"及澳大利亚和新西兰联合化学工程师"ExxonMobile Award",澳大利亚前 100 名最具影响力工程师等。

# 化学循环技术:俄亥俄州立大学铁基过程

徐迪凯　范良士*

俄亥俄州立大学 William G. Lowrie 化学与分子生物工程实验室,
哥伦布,俄亥俄,美国

**摘要:**近年来,在转化含碳燃料中,化学循环过程作为一种新颖的碳捕捉技术得到了广泛研究。化学循环过程用氧载体间接燃烧含碳燃料,原位捕捉 $CO_2$,而不是用传统的燃烧后捕捉技术。俄亥俄州立大学(OSU)开发了一种新颖的铁基化学循环技术,它能够最大限度地转化氧载体和燃料。热力学分析显示,当燃料完全转化,在对流移动床反应器中的氧载体的转化率远远高于其他反应器,而且通过水蒸气-铁法还产生了高纯度的氢气。

迄今为止,合成气化学循环(SCL)和煤直接化学循环(CDCL)技术已经在两个 25 $kW_{th}$的亚中试装置上证明完全成功。

## 一、引言

随着全球温室气体的排放,人们对气候变暖问题越来越关注。由于化学循环技术可以在消耗燃料时达到 $CO_2$ 的零排放[1],所以该技术已经得到大家的广泛研究。在化学循环过程中,一个固体氧载体,通常是金属氧化物,用来将含碳燃料在反应器中转化为一股含有二氧化碳($CO_2$)、水($H_2O$)、氢气($H_2$)和一氧化碳(CO)的气流,这样避免了燃料和空气的混合以及消除了之后的 $CO_2$ 的分离程序[2]。减少的氧载体可以通过两种方法再生,可以在单反应器中使金属直接被空气氧化,反应中放出的热量可以用于发电;也可以用双反应器,用蒸汽进行局部氧化和制氢气,用空气进行全氧化。

近年来,全世界研究了化学循环反应的各个方面,不同形式的氧载体和反应器都研究到了[2]。大部分氧载体的氧化物为氧化镍、氧化铁或氧化铜,作为活性

---

* 通讯作者,E-mail: fan. 1@ osu. edu。

组分并负载在惰性组分上，如 $Al_2O_3$。尽管镍基氧载体活性很好，但是它的成本过高而且容易中毒，并且会产生大量的 CO 需要后处理。铜基氧载体在高温下分解并释放氧气，因此有很好的活性，然而，由于铜的熔点低，导致其在流化床反应器中，容易凝聚和反流态化。为了克服上述缺点，在配方中采用低氧化铜负载比率，但牺牲了氧的携带量。铁基氧载体更加经济和环保。热力学显示，铁基氧载体比镍基氧载体有更低的 CO 产量。选择合适的载体有利于克服反应中的不足。循环流化床(CFB)被广泛地用做燃料反应器，因为它能处理大通量的固体物质并使气体、固体充分接触作用。然而，在流化床反应器中用铁基氧载体，由于热力学局限性，$Fe_2O_3$ 只能还原为 $Fe_3O_4$，相当于 11%的氧载体的能力被用到了。

俄亥俄州立大学(OSU)开发了一种独特的铁基化学循环技术，可以将固体或气体燃料全部转化为 $CO_2$ 和蒸汽。如图 1 所示，OSU 化学循环技术包含还原室、氧化室和燃烧室。主要化学反应如下：

还原室：$$C_xH_yO_z+Fe_2O_3 \longrightarrow CO_2+H_2O+Fe/FeO \quad (1)$$

氧化室：$$Fe/FeO+H_2O \longrightarrow Fe_3O_4+H_2 \quad (2)$$

燃烧室：$$Fe_3O_4+O_2 \longrightarrow Fe_2O_3+Q \quad (3)$$

总反应：$$C_xH_yO_z+H_2O+O_2 \longrightarrow CO_2+H_2O+H_2+Q \quad (4)$$

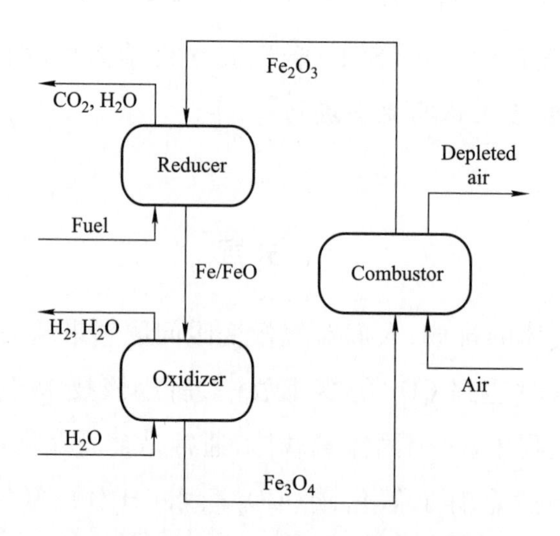


**图 1　俄亥俄州立大学化学循环反应图解**

气态产物气流离开还原室，冷凝掉蒸气后主要成分是 $CO_2$，冷凝蒸气被隔离或者被使用。同时，氧化室产生高纯度的 $H_2$，燃烧室用于加热。还原室和氧化室都是对流反应器，这种独特的对流反应器可以最大限度地转化燃料、蒸汽和氧载体[2-4]。

迄今为止，OSU 已经设计和建造了一个 25 $kW_{th}$ 合成气循环反应亚中试实验装置(SCL)，和一个 25 $kW_{th}$ 的煤直接化学循环亚中试实验装置(CDCL)。在上面进行了长达 850 h 的综合运转，并证明燃料得到了完全转化[5-8]。SCL 装置在合成气燃烧中实现了生成纯度大于 99.99% 的 $H_2$ 和生成纯 $CO_2$ 的目标。CDCL 装置被证明可以转化多种固体燃料，包括生物质、煤炭、冶金焦等。

## 二、氧载体的开发

在化学循环过程中，氧载体在反应器间流动，通过循环的氧化-还原反应在空气/水蒸气和燃料间进行氧的传送。因此，令人满意的氧载体需要在经过多次循环反应后仍保持较好的机械稳定性和化学活性，以及通过减少再生过程使其减少损失。广泛研究的氧载体材料包括硫酸钙、氧化镍、氧化铜、氧化铁和氧化锰。固体载体包括 $Al_2O_3$ 和 $MgAl_2O_4$，用于提高氧载体的机械强度和活性[9]。

OSU 还测试了 600 多种以不同金属氧化物、负载载体和合成方法为基础的氧载体材料。开发了一系列以 $Fe_2O_3$ 为基础的复合氧载体，发现其有良好的反应能力和循环能力。如图 2 所示，OSU 复合氧载体在超过 100 次氧化-还原循环后仍保持活性，而纯的 $Fe_2O_3$ 在几次循环后便丧失活性。与其他氧载体材料相比，铁基氧载体具有令人满意的热力学性质，氧化剂（$Fe_2O_3$）能够将燃料完全转化为 $CO_2$ 和 $H_2O$，还原剂（Fe/FeO）能够与蒸气反应生成 $H_2$。

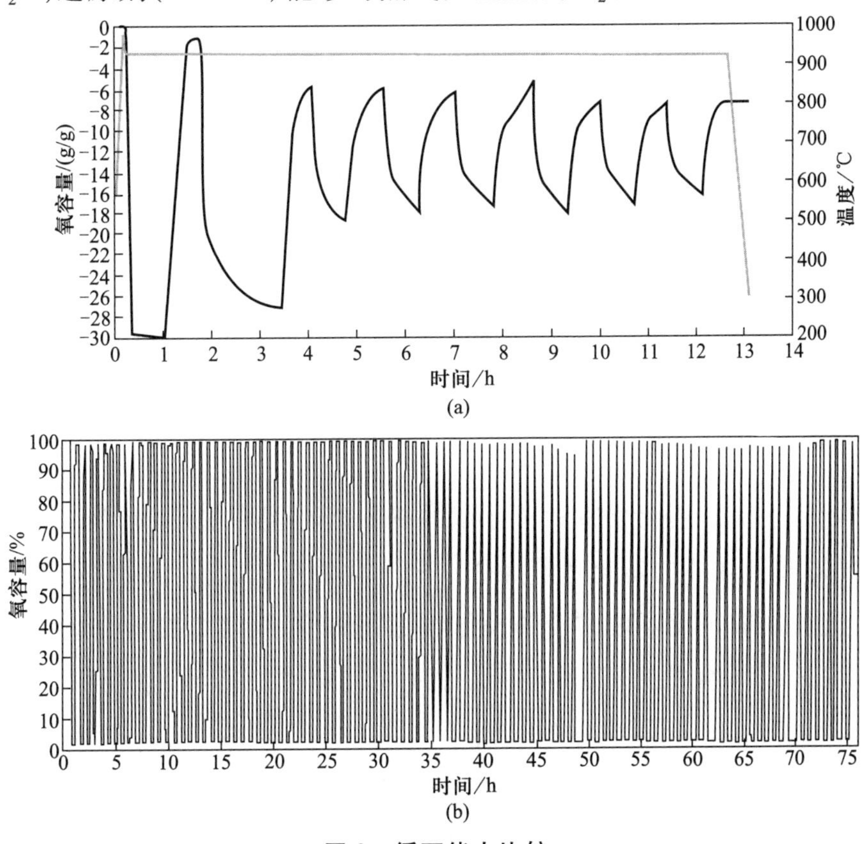


**图 2　循环能力比较**

(a)纯 $Fe_2O_3$；(b)OSU 复合氧载体

## 三、铁基化学循环反应器的设计与操作

到目前为止,多种化学循环反应示范装置已经被建造出来,全部反应时间超过 2000 h[10-13],绝大部分项目用的是流化床。OSU 化学循环技术采用独特的对流移动床还原室和氧化室,它能最大化地转化氧载体和全面地氧化燃料[14]。

从热力学的气体产物与固体产物达到化学平衡的角度来看,气固流化床反应器的表现与水平移动床反应器相一致。因此,还原室和氧化室的气固接触模式可分为两类:模式 1 流化床或平行移动床反应器;模式 2 对流移动床反应器[14,15]。每种模式中,都有特定的热力学局限性存在于氧载体和燃料转化中。

图 3 显示 $FeO_x$ 与 $H_2/H_2O$ 在 850℃反应的化学平衡。实线表示平衡状态:垂直的线表示气相介质和固相介质在一定的组成下的平衡状态。水平线表示一种气相与固相达到柔性组合的平衡状态。还原室在图的左下方,氧化室在图的右上方。

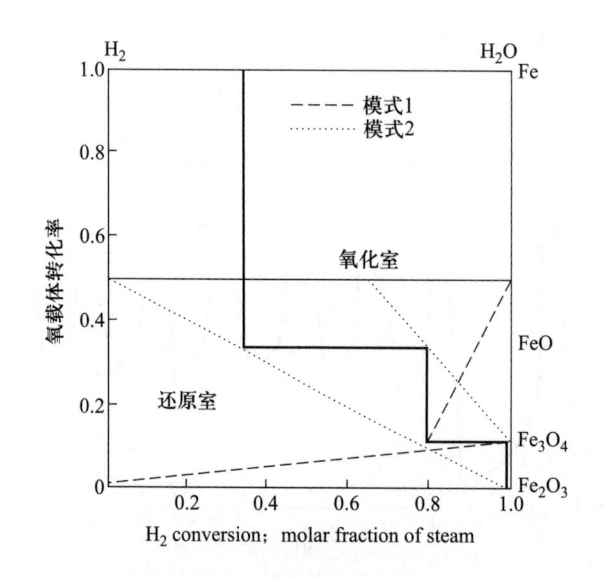


图 3 还原剂和氧化剂操作线

氧的物料平衡是用氧化室和还原室的操作线在图中显示的[14],沿着气/固流动方向,气体和氧载体的转化随着操作线的变化而变化。操作线的坡度与气体和氧载体的流量比成正比。模式 1 的操作线是正斜率,模式 2 的操作线是负斜率。

操作线的两种模式在图 3 中显示,虚线代表模式 1,点线代表模式 2。图 3 的还原室区,用模式 1 时,氧载体的转化率只有 11%,即由 $Fe_2O_3$ 生成 $Fe_3O_4$ 的过程。然而,在模式 2,氧载体的转化率可达 50%,$Fe_2O_3$ 被还原为 Fe 和 FeO 的混合物,因此,转化同样数量的燃料,模式 2 只需模式 1 的 20%的循环次数。而且,在模式 1 中,还原后的氧载体($Fe_3O_4$)不能在氧化室作用下与蒸汽反应生成氢气。

当模式 2 用在还原室，被还原的氧载体能够在氧化室作用下与蒸汽生成 $H_2$。氧化室的操作线在图 3 的氧化剂区域。得到同样的氧载体转化率（从 Fe/FeO 到 $Fe_3O_4$），模式 1 产生了 $H_2$ 和 $H_2O$ 的混合物，其中，$H_2$ 浓度占 20%，在模式 2 中，$H_2$ 浓度占 35%。因此，模式 2 的蒸汽流量和能耗都是更低的。如果氧化室的操作温度更低的话，蒸汽的转化率将大于 50%。

铁基化学循环反应系统，用对流移动床还原室和氧化室要优于用流化床和平行移动床。OSU 还设计建造了对流移动床以进一步研究化学循环反应系统。



## 四、铁基化学循环过程

### （一）合成气循环（SCL）过程

SCL 技术转化气体燃料（合成气或天然气）为电能和 $H_2$，并且无 $CO_2$ 排放。SCL 过程原理在图 4[2] 中显示，合成气（可由煤的气化得到）经过净化后可送到对流移动床的底部。从反应器中得到的高温气流的能量可以通过发电获得。过程中产生的氢气可以用作燃料电池或下游的化工生产。

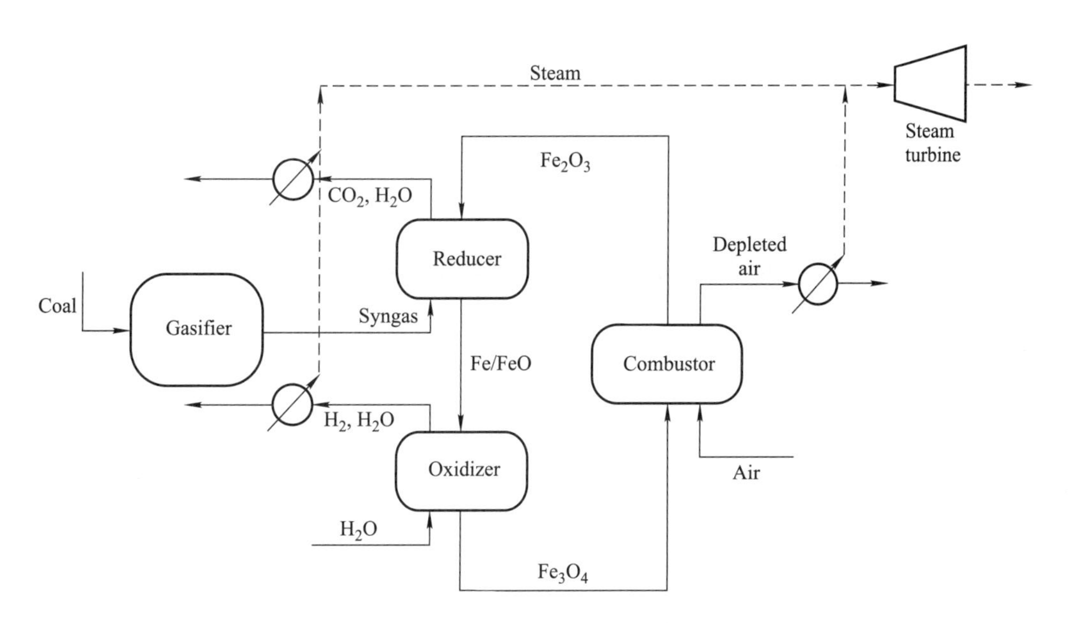


图 4　合成气循环系统图解

OSU 已经建造和运行了一个 25 $kW_{th}$ SCL 实验装置，它在合成气燃烧过程中能够得到高纯度的 $H_2$ 并完全捕捉碳[5,6]。一个反应连续 3 天都得到了稳定的高纯度 $CO_2$ 和 $H_2$，数据显示，合成气在还原室中几乎完全转化，在氧化室中得到纯度大于 99.99% 的 $H_2$[6]；也进行了燃烧甲烷的实验，甲烷的转化率大于 99.5%，实验装置的成功操作证明了 OSU 化学循环技术的可行性。

目前,OSU 在国家碳捕捉中心(NCCC)进行了 250 $kW_{th}$ 中试规模的加压 SCL 过程研究,此装置设计用 Kellogg Brown & Root (KBR) 燃气发生炉在 10 个大气压下燃烧合成气。SCL 中试装置已经做好,将于 2014 年年底进行实验。

### (二) 碳直接化学循环(CDCL)过程

OSU CDCL 技术能够直接将固体燃料如煤炭、生物质、冶金焦炭进行还原,不用煤气化炉。CDCL 过程有两级对流移动床和一个流化床燃烧室,固体粉末燃料引入还原室并加热除去挥发组分,还原室分为两段。气态的挥发物向上到减压器上方,被 $Fe_2O_3$ 充分氧化为 $CO_2$ 和 $H_2O$(图 5)。脱挥发分作用后的固体残渣和氧载体一同进入下层,并有少量的 $CO_2$ 和(或)$H_2O$ 被引入用来促进残渣气化[7,16]。

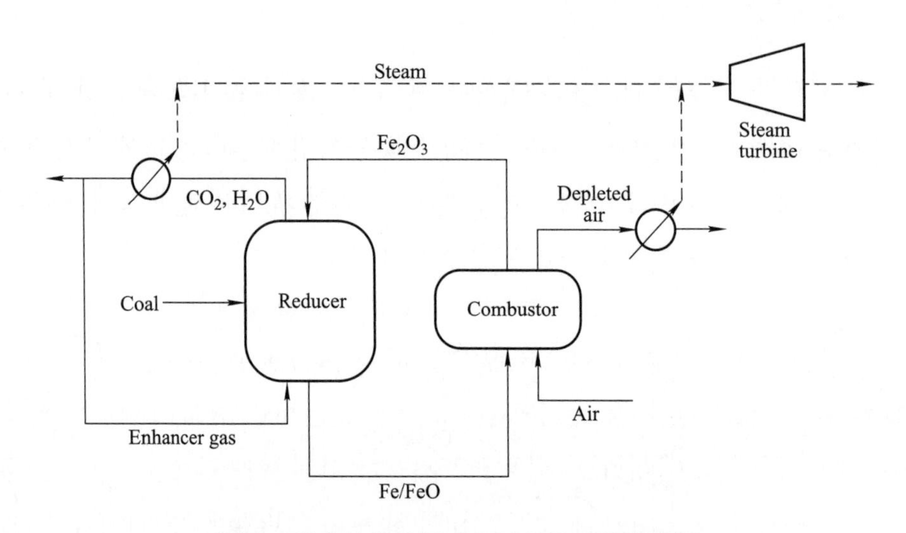


**图 5　碳直接化学循环(CDCL)过程流程图**

OSU 已经建立了一个 2.5 $kW_{th}$ 实验室规模的移动床反应器和一个 25 $kW_{th}$ 亚中试规模的联合装置去研究转化冶金焦、褐煤、次烟煤、烟煤、无烟煤、木质生物质[7,8,17-19]。转化实验超过 300 h,各种挥发分和固体燃料都在这个 2.5 $kW_{th}$ 的实验室规模的装置上试验过。

25 $kW_{th}$ 亚中试规模的装置完全采用非机械结构进行气固流的控制。实验超过 550 h。最长的连续操作长达 200 h,它也是目前为止固体燃料化学循环过程持续时间最长的[9]。整个运行过程很稳定,气体固体流量都得到了很好的控制。而且其转化 PRB 煤和褐煤的能力也得到了验证。这两种煤在亚中试的实验装置中的转化率大于 96%,从还原室中出来的 $CO_2$ 纯度在 99.4% 到 99.8%之间,副产物 $CH_4$ 和 CO 的浓度低于 0.5%。结果显示,还原室可以完全氧化固体燃料,少量的 $CO_2$ 在燃烧室的出口被检测到,证明有极少量的碳从还原室传送到燃烧室。

### （三）将化学循环技术与其他技术集成

OSU 的化学循环技术有高度的灵活性来接收不同的燃料并输出想要的能源，如电能、氢气和液体燃料。图 6 显示了整合 OSU SCL 技术和传统的煤制液体（CTL）过程[2,15]，SCL 可以结合到现有的 CTL 系统中，提高反应过程效率并减少 $CO_2$ 排放。

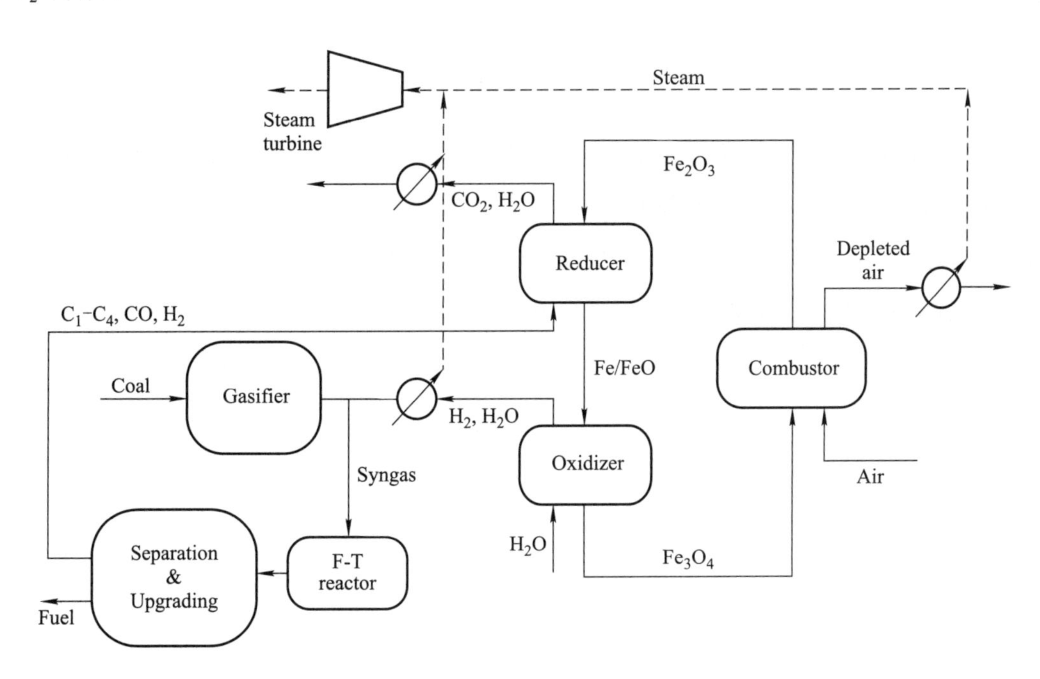


**图 6　SCL 和 CTL 技术的集成**

在此集成中，通过调整 SCL 的氧化室生成的 $H_2$ 的量，来使得合成气的 $H_2$：CO 达到一个合适比例，从而有助于进行费托合成过程。费托合成出来的轻烃和提炼出的产品可作为 SCL 过程的燃料，并转化为准备隔离的 $CO_2$ 气流。此气流产生于费托合成反应，可以很方便地作为 SCL 氧化室的原料。因此，SCL 过程替代了传统的 CTL 过程水煤气转换和 $CO_2$ 移除的过程。美国能源部一份独立的分析显示，与传统的 CTL 过程相比，集成方法可以使液体产量提高 10%，减排 $CO_2$ 19%[20]。

## 五、化学循环技术的技术经济分析

对 OSU 化学循环技术一系列的技术经济分析表明，其技术比现有碳捕捉技术效率更高、成本更低。图 4 显示 SCL 技术比传统的煤气化制 $H_2$ 效率提高 5.5%，并且有 90% 的 $CO_2$ 捕捉率。如果 $H_2$ 能够以 60% 的效率转化为电能，那么平均化用电成本将降低 10%[21]。

CDCL 过程发电优于传统的碳捕捉粉煤(PC)发电厂,一个 550 $MW_e$ 的发电厂,CDCL 过程有 100%的 $CO_2$ 捕捉率,比粉煤发电厂用 MEA 捕捉 90% $CO_2$ 的效率提高 23.5%。

用电量只比粉煤发电厂不加碳捕捉装置时提高 33%。而当粉煤发电厂通过 MEA 捕捉 90%的 $CO_2$ 时,其能耗将提高 71%[22]。因此 CDCL 过程可以达到美国能源部的要求,即电能消耗减少 35%,碳捕捉提高到 90%。

## 六、结论

OSU 开发了独特的铁基化学循环过程,采用对流移动床装置,能够高效地转化含碳燃料并实现 $CO_2$ 的零排放。合成和测试了超过 600 种氧载体。目前的铁基复合氧载体能在超过 100 次循环后依然保持活性。氧载体的性能在实验室规模和亚中试规模的实验装置中进一步得到了印证。根据铁基氧载体的热力学性质,OSU 设计了独特的对流移动床还原室和氧化室。它能最大化地转化氧载体并保证燃料全部转化。相对于流化床反应器,固体在对流移动床的循环率要低 80%,$Fe_2O_3$ 可以被还原为低价态的氧化态,使氢在氧化室生成。因此,化学循环技术可以和很多技术集成,比如和 CTL 结合达到更高的能源效率。

OSU 化学循环技术已经在两个 25 $kW_{th}$的亚中试反应器上证明成功,SCL 亚中试反应器可得到纯度大于 99.99% 的 $H_2$ 并在合成气燃烧中完全捕捉 $CO_2$。一个 250 $kW_{th}$的加压中式 SCL 装置已经在国家碳捕捉中心建造并在 2014 年年底运行。CDCL 技术运用了二级对流移动床还原室和无机械结构控制气固流向、流速。气体、固体流量被证明长时间运行稳定。进行了长达 550 h 的运转,所包含的燃料有焦、煤、生物质等。

(感谢王威廉在技术和编辑方面的帮助。)

(此工作是由俄亥俄州立大学的 C 约翰伊斯顿基金资助的。)

## 参考文献

[1] Figueroa J D, Fout T, Plasynski S, et al. Advances in $CO_2$ capture technology—The US Department of Energy's Carbon Sequestration Program[J]. International Journal of Greenhouse Gas Control, 2008, 2(1): 9-20.

[2] Fan L S. Chemical looping systems for fossil energy conversions[M]. New York: Wiley, 2010.

[3] Li F, Fan L S. Clean coal conversion processes-progress and challenges[J]. Energy & Environmental Science, 2008, 1(2): 248-267.

[4] Thomas T J, Fan L S, Gupta P, et al. Combustion looping using composite oxygen carriers: U.S. Patent 7,767,191[P]. 2010-8-3.

[5] Sridhar D, Tong A, Kim H, et al. Syngas chemical looping process: Design and construction of a 25 $kW_{th}$

subpilot unit[J].Energy & Fuels,2012,26(4): 2292-2302.

[6] Tong A,Sridhar D,Sun Z,et al.Continuous high purity hydrogen generation from a syngas chemical looping $25kW_{th}$ sub-pilot unit with 100% carbon capture[J].Fuel,2012,103:495-505.

[7] Kim H R,Wang D,Zeng L,et al.Coal direct chemical looping combustion process: Design and operation of a $25-kW_{th}$ sub-pilot unit[J].Fuel,2013,108:370-384.

[8] Bayham S C,Kim H R,Wang D,et al.Iron-based coal direct chemical looping combustion process: 200-h continuous operation of a $25-kW_{th}$ subpilot unit[J].Energy & Fuels,2013,27(3): 1347-1356.

[9] Fan L S,Li F.Chemical looping technology and its fossil energy conversion applications[J].Industrial & Engineering Chemistry Research,2010,49(21): 10200-10211.

[10] Adanez J, Abad A, Garcia-Labiano F, et al. Progress in chemical-looping combustion and reforming technologies[J].Progress in Energy and Combustion Science,2012,38(2): 215-282.

[11] Andrus H E,Burns G,Chiu J H,et al.Hybrid combustion-gasification chemical looping coal power technology development phase III final report, Alstom Power Inc[R].PPL-08-CT-25, Contract DE-FC26-03NT41866,US Department of Energy,National Energy Technology Laboratory,2006.

[12] Kolbitsch P,Bolhar-Nordenkampf J,Pröll T,et al.Comparison of two Ni-based oxygen carriers for chemical looping combustion of natural gas in 140 kW continuous looping operation[J].Industrial & Engineering Chemistry Research,2009,48(11): 5542-5547.

[13] Shulman A,Linderholm C,Mattisson T,et al.High reactivity and mechanical durability of $NiO/NiAl_2O_4$ and $NiO/NiAl_2O_4/MgAl_2O_4$ oxygen carrier particles used for more than 1000 h in a 10 kW CLC reactor[J].Industrial & Engineering Chemistry Research,2009,48(15): 7400-7405.

[14] Li F,Zeng L,Velazquez-Vargas L G,et al.Syngas chemical looping gasification process: Bench-scale studies and reactor simulations[J].AIChE Journal,2010,56(8): 2186-2199.

[15] Fan L S, Zeng L, Wang W, et al. Chemical looping processes for $CO_2$ capture and carbonaceous fuel conversion-prospect and opportunity[J].Energy & Environmental Science,2012,5(6): 7254-7280.

[16] Li F,Zeng L,Fan L S.Biomass direct chemical looping process: Process simulation[J].Fuel,2010,89(12): 3773-3784.

[17] Luo S,Bayham S,Zeng L,et al.Conversion of metallurgical coke and coal using a coal direct chemical looping (CDCL) moving bed reactor[J].Applied Energy,2014,118(1):300-308.

[18] Luo S,Majumder A,Chung E,et al.Conversion of woody biomass materials by chemical looping process: Kinetics,light tar cracking, and moving bed reactor behavior[J].Industrial & Engineering Chemistry Research,2013,52(39): 14116-14124.

[19] Zeng L,He F,Li F,et al.Coal-direct chemical looping gasification for hydrogen production: Reactor modeling and process simulation[J].Energy & Fuels,2012,26(6): 3680-3690.

[20] Gray D,Klara J,Tomlinson G,et al.Chemical-looping process in a coal-to-liquids configuration: Independent assessment of the potential of chemical-looping in the context of a Fischer-Tropsch plant[R]//Report No.: DOE/NETL-2008/1307.Contract No.: NBCH-C-020039.Sponsored by the US Department of Energy. Pittsburgh P A: National Energy Technology Laboratory,2007.

[21] Li F, Zeng L, Fan L S. Techno-economic analysis of coal-based hydrogen and electricity cogeneration

processes with $CO_2$ capture[J].Industrial & Engineering Chemistry Research,2010,49(21): 11018-11028.

[22] Connell D,Dunkerley M.Techno-economic analysis of a coal direct chemical looping power plant with carbon dioxide capture[C]//Proceedings of the 37th International Technical Conference on Clean Coal and Fuel Systems,June 3-7,2012.Clearwater,USA,2012:29-40.

**范良士**　美国工程院院士，中国工程院、澳大利亚科学技术与工程学院和墨西哥科学院外籍院士。美国俄亥俄州立大学化学与分子生物工程学院杰出教授和C. John Easto教授。他的研究领域是流体学、粉末技术和多相反应工程。2008年被美国化学工程学会评为“当代百名工程师”。

# 基于光催化的清洁能源与环境新技术

付贤智

福州大学

能源与环境光催化国家重点实验室

国家环境光催化工程技术研究中心,福建福州

**摘要**:能源短缺和环境污染是现代社会的重要问题。光催化技术能够利用日光催化许多相关的化学反应,如水裂解产氢、二氧化碳还原和有机污染物的降解,因此是未来生产清洁能源和环境修复的理想方法。迄今为止,光催化技术的实际应用局限于光催化过程的低量子效率和日光的低利用效率。为解决这些问题,科学家们正在进行大量的基础和应用研究,主要集中在新的光催化剂的设计和合成,结晶相/面的调整,异质结和助催化剂的制作,光催化剂的带结构工程,光催化的反应机理等。报告将涵盖这些方面的主要研究进展,并伴有环境光催化技术的一些应用实例,同时也对光催化技术的一些未来前景进行了探讨。

我报告的题目是"基于光催化的清洁能源与环境新技术",包括研究背景与光催化简介、光催化最新研究进展、光催化技术应用示例和展望四个方面的内容。首先简要介绍一下光催化的研究背景。能源和环境问题是21世纪人类共同面临和亟待解决的重大问题,今天上午很多大会报告也都涉及了这个问题。从能源来看,全球初级能源消耗在不断增加,但作为主要能源的煤炭、石油和天然气等化石能源由于大量长期开采在逐渐枯竭。从环境来看,全球的环境污染在日益恶化,无论是土壤污染、大气污染还是水污染,形势都非常严峻,同时由于化石能源的大量使用,近年来二氧化碳的排放量不断增加,温室效应使得全球气温上升,导致冰川逐渐消失、海平面上升、水灾频发。

面对严峻的能源与环境问题,人们把目光转向可再生能源的开发与利用,其中,太阳能尤为引人关注,因为它是取之不尽、用之不竭的清洁能源。据统计,每年到达地球表面的太阳能相当于130万亿吨煤,一小时的太阳辐射能量够地球使

用一年。目前太阳能的主要利用方式有光热、光电和光化学转化,在光化学转化中,光催化是近年发展起来的,可以利用太阳能进行环境净化、能源转化和二氧化碳还原的新技术,通过光催化剂的作用,把水分解成氢气和氧气,将太阳能转化为洁净的氢能源,也可以通过光催化反应把空气和水中的有机污染物分解等。

从目前的研究情况来看,环境光催化已经开始步入实际应用,能源光催化和 $CO_2$ 光催化转化还处于实验室研究阶段。光催化过程本质上是光诱导的氧化-还原反应过程,包括光激发形成光生电子-空穴对,光生载流子的分离、迁移和重新复合,催化剂表面的氧化-还原反应等三个基本过程。作为环保新技术,光催化技术具有可以直接利用太阳能净化环境,在室温下彻底降解污染物,有效灭杀细菌、病毒,安全、无二次污染,广谱、长效等特征。作为能源新技术,光催化技术具有可以直接把太阳能转化为高能量密度的化学能,零碳或碳循环过程,转化过程条件温和,安全、无二次污染等特征。

目前以二氧化钛为基础的半导体光催化,要实现大规模的实际应用还有一系列的科学问题和技术问题需要解决。由于二氧化钛的结构和性质决定了形成的光生载流子容易重新复合,导致光催化过程量子效率偏低;另外,由于它是宽带隙半导体,带隙宽度是3.2 eV,只能吸收利用太阳光中少量的紫外线,而大部分的可见光得不到有效利用,导致光催化过程的太阳能利用率偏低。另外,由于光催化反应涉及气-固-液-光多相反应,所以反应系统设计等工程化技术更加复杂。以上问题的解决有赖于光催化剂构-效关系及调控、光催化过程的作用机理、提高光催化过程效率的途径等基本科学问题的解决。因此如何从理论和应用上解决这些问题是当前和今后一个时期国际光催化领域的研究前沿与热点。

针对上述重大科学与技术问题,近年来开展了系统深入研究,在可见光诱导的新型高效光催化剂、多相光催化反应过程机理、提高光催化过程效率新方法、拓展光催化技术应用新领域等方面均取得重要进展,光催化成为当前国际研究的热点之一,有关光催化的研究论文数量近年来快速增加,并且重要的研究成果经常能够在《自然》、《科学》等高水平刊物上得到及时报道。从中国的光催化研究来看,2000年开始我国光催化研究实现了跨越式发展,根据国外学者统计,目前我国在国际上发表光催化论文的数量已经居于世界第一位,光催化研究领域已经成为我国具有国际竞争力的学科领域之一。

下面结合我们的工作,简要介绍一下光催化领域的最新研究进展情况。在众多的研究进展中,新型光催化剂的研究进展尤其突出,因为光催化的核心是光催化材料,很多问题主要是源于光催化材料,所以大家把很多力量都放在了设计、研究新型的光催化剂上。采用表面修饰、半导体复合、分子设计合成、离子掺杂、固溶体形成、量子尺寸效应、晶面/相调控、助催化剂/异质结等多种方法和手段,对催化剂的能带、结构、组成、表面进行调控,设计并制备出一系列具有高效、高稳定性和可见光诱导性能的新型光催化剂,大大拓展了光催化剂的多元性和应用可选择性。新型光催化剂大体上可以分为两大类:一类是对二氧化钛进行改性的 $TiO_2$ 基新型光催化剂;一类是不含二氧化钛的非 $TiO_2$ 新型光催化剂。下面举一些新型光催化剂研究进展的例子。对于 $TiO_2$ 基新型光催化剂,第一个例子是非金属元素掺杂型二氧化钛光催化剂,Asahi 等报道了以 N 为代表的非金属掺杂 $TiO_2$,证实其在可见光照射下具有很好的光催化降解有机物活性和超亲水性,同时又不会削弱 $TiO_2$ 在紫外线下的光催化性能;第二个例子是表面晶格缺陷型二氧化钛光催化剂,Mao 等对 $TiO_2$ 进行氢化处理,使其表面产生大量晶格缺陷,通过表面调控,有效降低了禁带宽度,使其光吸收发生明显红移,颜色从白色变为黑色,具有可见光光催化活性;第三个例子是晶面调控型二氧化钛光催化剂,Gaoqing (Max) Lu 和成会明等以 HF 为封端剂,在水热条件下控制 $TiO_2$ 的晶面生长,制备出分别以{001}、{101}、{010}晶面为主的锐钛矿型 $TiO_2$ 单晶,研究了晶面表面原子结构和电子结构与光催化活性间的关系,为实现高效光催化剂可控制备提供了新方法。对于非 $TiO_2$ 新型光催化剂,第一个例子是固溶体型可见光光催化剂,Domen 等采用高温氮化的方法,制备出具有良好可见光响应和较高光解水活性的氮(氧)化物固溶体光催化剂,如 GaN-ZnO、GeN-ZnO 等,其中,GaN-ZnO 在 420~440 nm 下的量子效率达到 2.5%;第二个例子是非金属聚合物型光催化剂,Xinchen Wang 等最近研究发现,石墨相氮化碳($g-C_3N_4$)具有可见光光催化活性,并将其作为光催化剂引入光催化研究领域,采用共聚合法、模板法、金属 Pt 改性和催化剂表面异质结构的构筑等方法合成出一系列新型高效氮化碳光催化剂,拓展了光催化研究特别是光催化材料研究的新方向;第三个例子是金属有机框架结构型光催化剂,Zhaohui Li 等用氨基对苯二甲酸取代紫外线响应 MIL-125(Ti)中的链接分子对苯二甲酸,得到了具有相似结构且可见光响应的 MOF 材料 NH2-MIL-125(Ti),首次将 MOF 应用于可见光光催化还原 $CO_2$。

另外光催化反应机理的研究也取得了重要进展,初步阐明了若干重要光催化过程的反应机理和作用本质,并对影响光催化反应速率的关键因素有了进一步认识。举几个例子,比如石墨相氮化碳的有机物光催化选择性氧化机理,Xinchen Wang 等的研究表明,$g-C_3N_4$ 通过光生电子活化分子氧,形成超氧自由基负离子,实现对有机物的光催化选择性氧化;陈接胜等发展了 $Zn^+$改性的 ZSM-5 分子筛光催化剂,对 C—H 具有很好的光催化活化作用,能够实现甲烷的光催化偶联反应,生成乙烷和氢气,并证实其活性中心为 ZSM 分子筛骨架中的 $Zn^+$,提出了甲烷光催化偶联反应机理；多相光催化过程的酸性作用机理研究取得进展，我们的研究工作表明，增强催化剂表面酸性可有效抑制光生电子-空穴重新复合，使光催化活性提高。$SO_4^{2-}/TiO_2$ 固体超强酸表面的 Brønsted 酸中心是光催化活性中心，赵进才等的进一步研究工作表明，Brønsted 酸的作用在于光助分解表面 $\eta^2(Ti-O_2)$ 过氧物种，加速更新 $TiO_2$ 催化剂表面。

近年来光催化应用研究也取得了重要进展，研制开发出多种提高光催化过程效率和解决光催化应用工程化关键问题的新技术、新方法和新装置，光催化应用领域不断拓展，在大气净化、室内空气净化、土壤净化、饮用水净化、工业污水处理、防污闪高压绝缘子、自清洁建筑材料、自清洁抗雾玻璃、医疗卫生等众多方面都有成功的应用实例。

综上所述，近年来在光催化领域无论是基础研究还是应用研究都有长足进步和发展，为光催化技术将来在能源和环境领域能够大规模实际应用奠定了很好的基础。从发展趋势来看，基础研究方面，随着基础研究的深入开展，对光催化反应过程微观机理和调控规律等关键科学问题的认识将不断深化，在此基础上将建立起基于分子水平的、具有普适性的光催化作用新理论，在新理论的指导下将设计并制备出性能更加优异并可以实际应用的多种新型高效光催化剂；应用研究方面，在提高光催化过程效率的应用关键技术方面将会不断突破，同时随着应用研究的不断深入，光催化技术新的应用领域将不断拓展；光催化产业发展方面，随着新型光催化材料和应用关键技术的不断突破，将有力促进光催化环保产业的快速发展，利用太阳能光催化分解水制氢和人工光合成的产业化将成为可能。如果这些都能够实现大规模的产业化，那么最终将形成一个以光催化技术为基础的巨大的高新技术产业链。

从理论上来讲，光催化确实是解决环境与能源问题的理想途径之一，但从目前的情况来看理想尚未完全变成现实，刚才提到的重大科技问题只解决了苗头并且进展很好，但大多还没有完全解决。所以，我们作为在这个领域开展工作的研究者还应该加倍努力，争取能够做得更好。

最后用一句话来结束今天的报告，太阳给世界带来光明和温暖，光催化为人类创造洁净和健康。

谢谢大家！

**付贤智**　中国工程院院士，现任福州大学教授、国家环境光催化工程技术研究中心主任、能源与环境光催化国家重点实验室主任。长期从事光催化基础与应用研究，主要研究方向为光催化剂的设计与制备、光催化反应机理、光催化反应动力学和光催化反应器设计。研制出一系列新型高效光催化剂，开发出提高光催化过程效率的多种新技术、新方法和新装置，并将光催化技术应用于环保、建材、军工、电力等领域，研制开发的多项光催化产品实现了产业化。迄今在国内外重要学术刊物上发表研究论文300多篇，授权的国家专利40余项，研究成果先后获得国家科技进步奖二等奖1项、军队科技进步奖一等奖1项、省部级科技进步奖一等奖3项。

# 可诱导组织再生的生物材料
# ——生物材料发展的新纪元

张兴栋

四川大学国家生物医学材料工程技术研究中心，四川成都

**摘要**：用于诱导组织再生的生物材料是指植入时，能够引发生物反应和修复再生受损组织或器官的生物材料或植入装置。通过材料、组织工程产品，以及控释的媒介及能够使组织再生的药物和基因系统的优化设计，这些生物材料能够诱导组织再生。传统的无生命的生物材料已经被证明是非常成功的，但临床实践已证明它们不能完全满足临床要求，如功能性、寿命等。因此，传统的无生命的生物材料时代正在终结，而生物材料科学及其产业正在经历一个革命性的变化。重要的是，能够诱导组织再生的生物材料已经显著地成为生物材料发展的前沿和方向，在不久的将来，将成长为生物材料产业的一个重要部分。生物材料诱导组织再生的非常关键的原理是无生命的材料可以诱导活组织或器官的再生。这一理念，从非生命到生命的演变，是传统智慧的一个突破。在报告中，我将介绍我们的突破性的发现，也就是无生命的生物材料诱导活的骨组织的再生。随后讨论诱导机理。最后，从组织诱导的生物材料来看，将给出非骨组织诱导的生物材料潜力的观点。

**张兴栋**　中国工程院院士，美国国家工程院外籍院士。现任四川大学教授，兼任中国生物材料学会理事长、国际组织工程与再生医学学会大陆（亚太）理事会理事、全国医疗器械生物学评价标准化技术委员会主任委员、全国口腔材料和器械设备标准化技术委员会主任委员等。1983年开始生物医用材料，特别是新一代肌肉骨骼系统修复材料及植入器械研究，获国家自然科学奖和国家科技进步奖二等奖各1项，省部级科技进步奖一、二等奖7项，国际奖两项，SCI收录论文350余篇，编著（含合作）图书10部，获国家食品药品监督管理局生产注册证6项。

# 玉米淀粉、番茄酱与汽车零配件：半固态加工技术综述

**Helen Valerie Atkinson**

莱斯特大学，莱斯特，英国

**摘要**：20 世纪 70 年代初，半固态金属的触变行为首次被 Flemings 和他在麻省理工学院的研究小组发现。当合金在凝固过程中搅拌，然后再加热成半固体状态时，它表现出令人惊讶的行为，将其放置在均匀流动的重机油上，进行切断加工时像加工固体一样。此行为（非常类似于番茄酱）被应用于一系列半固体加工技术，包括触变成型、触变锻造、触变铸造、流变铸造和流变成型。触变成型（基于喷射铸造法）被许多公司使用，尤其是在日本和美国生产的镁合金部件，如便携式计算机和摄像机，但它不适合于铝合金。主要由触变铸造和触变成型技术生产的铝合金已被广泛应用于汽车。在过去的几年中，调查发现中国、印度和其他国家对半固态加工的许多新的想法的兴趣大大增加。

**Helen Valerie Atkinson** 英国皇家工程院院士，英国皇家工程院副主席，莱斯特大学工程学院院长，材料力学研究组织主席。获得剑桥大学学士学位，帝国理工大学博士学位，比利时列日大学的荣誉博士学位。2011 年，因为她在材料科学与工程领域的突出贡献，获得了由中国科学院金属研究所和沈阳材料科学国家实验室颁发的“李薰讲座奖”。2013 年被聘为有色金属研究员客座教授。她是第一位大会工程教授委员会的女主席，并由于在科学、工程和技术领域的杰出贡献，2014 年获得“英帝国统帅奖”。

# 化学与材料工程领域未来产品与工艺研发的挑战

**K. V. Raghavan**

印度化学技术研究所,海得拉巴,泰伦加纳,印度

**摘要**:从根本上说,化学和材料工程已经从量子和固态化学、电子物理学、应用力学、数学和生物学衍生发展成为先进的科学工程。它们未来的产品/工艺创新在很大程度上取决于对其在分子水平上的转换的理解,该分子水平涉及分层组织上微米或纳米水平的尺寸以及化学键的形成的科学尺度。分子运动的多尺度模拟,操控单分子或多分子去创建具有较高的反应活性和可对分子实体及其性质进行试调控的微观结构,在这些行业的新产品设计开发上给人们提供了振奋人心的机会。

微生物生存系统的工程学认识对设计和构造具有消费者效用及对人体和动物系统有高生物活性的生物分子实体是极其重要的。类似的方法吸引了人们对新的生物材料实体的发现和开发的关注。利用公认的工程概念、解释复杂的生物转化、发现新的生物炼制的方法为石化和特殊化工领域提供范围广泛的环保型化工产品,仍然是一个挑战。从清洁技术考虑,利用自我优化化学系统生产一个理想的产品对新的治疗药物的催化体系和超纯物质的发展来说是一个非常好的选择。在化学和相关领域,催化与提供多种工艺技术的化学反应工程有很紧密的联系。对多相催化剂实体微环境的修饰,催化剂孔道内分子间相互作用的模拟以及反应器内部的流体动力学的计算流体动力学(CFD)模拟有助于实现商业化工过程系统中高特异性和反应速率。

在此,我将证明上述及其他工程技术的进步通过在典型工业制造工艺约束的影响下预测和改善它们的性能已经改变了化学和材料工程领域的局面。

我具体介绍一下化学与材料工程的未来产品、工艺、研发挑战,以及如何来为这样一种研发去找一个好的环境。

首先谈到化学与材料，这里指的是研究一些化学反应，一方面要有相应的化学科学知识，还要有一些材料科学知识，如果你要研究生物材料的话，那还必须要研究生物科学。在我们谈产品发展的过程时，这个发展过程应该是一种清洁的过程，这也是我们现在所热衷讨论的一个话题。提到化学工程，化学工程现在变得越来越多学科、跨学科，尤其是我们现在要研究的内容越来越多，这一点与 20 世纪五六十年代不一样，那时的化学工程与现在的化学工程完全不是一回事。目前的化学工程必须要对整个过程有全面的、跨学科的了解，对于整个过程都必须有非常深入的认知。

要研究的东西越来越多，包括分子的振动，现在也越来越跨学科，要有创新技术和发展的新产品。对于科学工程技术来说，每一样东西都有一些核心知识和核心技术，同时，我们也必须有其他的临近学科的知识，包括化学、物理和生物科学，一旦有了这方面的基础知识，你的研究就可以跨学科，而且还可以影响到整个过程。

我们在化学工程领域已经做了很多的研究，这是多方面的研究。许多研究，特别是涉及一些新知识的研究是在微观层面进行的。在未来，包括流动动力学、设备设计都将应用到现代化学工程与技术的研究中。因此，现代化学工程的研究趋势将由宏观层面向纳米级的微观层面发展。

谈谈化学工程和材料工程，化学工程与材料工程之间的边界一开始是不清楚的，有一些人认为化学工程是应用的化学，材料工程是应用物理学。实际上化学工程给了我们一个非常好的机会，使材料科学能够发现并且制造一些典型的产品，包括纳米复合材料以及其他复合材料。我们今天谈的这个分子工具用于材料合成，包括相应的纳米技术，这些纳米技术帮助我们对基因进行更好的研究，有助于医学在相应领域的进展。现在还有一些分子的模型，它能达到非常精确的水平，这也有赖于材料工程和化学工程的发展。生物学研究相对于化学工程研究而言，拥有极强的非线性和不可预测性。实际上很多化学过程同时也是生物过程，如酶的反应可以用米-门二氏动力学方程表示，反馈和前馈控制理论经过适当的修改也可以应用到生物学系统研究中。遗传学、基因组学、蛋白质组学等领域的最新研究仅用化学工程学是很难完全理解的，但如果应用到生物学的研究中，就能更好地理解这些反应的方程式了。现在，迅速发展的基因领域的研究也越来越多地应用于化学工程的研究中。

芯片在生命系统工程领域的应用越来越受到世界的瞩目。这种系统能够帮助我们更好地进行医学方面的发现，可以帮助我们更好地研究人类细胞和人类组织，对于人体器官包括肺、肾的运转进行一些模拟。在芯片上能对这些活体——

人体相应的系统进行工程的研究,同时这些技术也处于不断的竞争过程中。对于生物技术来说,现在的发展变得越来越复杂,当然其成果也越来越多。

对分子层面构效关系的理解,对于新的高活性化学实体和材料的研究有重要作用。实际上,新化学分子的发现、新催化剂的发现和新材料的发现这三个研究领域之间是有一些相同点的。在新化学品和人体学研究间存在着一些联系。人体环境需要进行模拟,催化剂作为一种微观介质,需要在特定的产生环境下进行精确测量,反应环境需要在 M1 和 M2 间建立。但人们对于这些研究的实践很有限。

这里有一些例子,首先是分子模型,了解这些分子的活性并将其选择出来进行检测是非常重要的。同时我们还需要知道哪种能量会引起分子变化从而导致疾病,需要通过它们的化学结构和生物结构信息来确定选择的分子模型是否有效。这是一个综合的平衡的选择,我们有许多候选者,但我们只能选我们需要的那个分子模型。就像有一个微型的化工厂,我们拥有原材料和最终产品,通过不同的途径我们可以合成超过 100 种材料。

还想提到的在模型方面很重要的一点,就是在催化剂里面发生了什么样的过程,这是六七年之前我们做的一个实验,实验发现有一个线性的过程,为了实现这个过程需要一些能量,这些能量在这个体系当中能够发挥作用,我想在这里提到的一点就是,不仅要看这个催化剂的反应,同时要看反应的过程发生了什么。现在有一个环境友好型化工过程的新概念被提了出来,它包括反应的分离、催化剂反应设计、外部刺激和自动平衡分子的运用。Trost 于 1991 年提出的原子经济的概念及 Anastas 和 Warner 博士于 1998 年提出的 E-因子概念,为环境友好型化工过程这个新概念的形成奠定了基础。

另一个很重要的技术是交叉学科的催化剂。多功能和交叉学科的研究是非常复杂的,不同领域的研究人员需要共同合作。我想给大家讲一个很有趣的发展,就是关于这个吸收,我们需要选择正确的符合目的、要求的流程,这个体系是很难做模型的,但是可以变得更加具体。我们看到有蒸馏、吸收和再吸收的过程,这是一个完整的过程,在化工领域可以看到很多这样的例子。下一步工作是在催化剂粒子周围创造一个导电的微环境。活性位置的生成物的定位是取得有效成果的关键因素。创造这个催化剂微环境的可选方法包括同类催化剂的异质化,新阶段的创造,微型笼子、微腔和微型结构的创造,液-液分散物和混合压缩金属分子筛的创造。

这是一个固体酸催化硝化反应,我们研究的一个目标就是如何实现催化剂效用的最大化,在这个反应蒸馏当中我们要确保环境的友好性,要选择正确的催化剂，实现它的经济性。这就是我们使用的一个体系，有一些分散的有机物，还有液体膜，这个地方装的是水，在膜的下面放催化剂，催化剂得到了过滤，然后我们就加强压力，这个过程就开始了。所以这就是一种蒸汽变成液体、液体变成固体的过程。我刚才还提到可以利用一些外部因素来提高效率，采用一种外界的媒介来做到这一点，应该采用什么样的媒介呢？可以有一些超液体，还有一种伪液体，这就是被吸收的分子发生了化学变化，还有微吸收，它涵盖了一些微粒、小的颗粒，还有小的液体滴，通过这样的方式都可以提高效率。

我们还可以用一些额外的力量来加大反应的力度，比如说微波、超声和膜等，这都是可以采用的加速器，可以为反应加强动力，现在我们有很多这样的反应器。想提一下这个膜，它可以通过渗透来实现目标，如果用这个传统的方法可能需要 9 h 才能完成，而且产出率只有 93%，而现在采用新型的方式可以达到 80%多的产出率，同时只需要 0.01 kW · h 的电，而刚才的方法需要 9 kW · h 的电。大家可以看到这个效率程度，我们需要正确地选择，使反应的体系更加清洁、更加高效。所以未来会是不断自我改变、自我改善的过程，这将给我们的清洁化学生产打开一个新的时代，我们会生产出更加清洁、纯度更高的化工制品。

自我组织是一个非常有趣的概念。这个概念之前被提出，大家可以看到，在右边是一群鸟，这些鸟在飞行的时候会自动组成一个有效的组织，而很有趣的一点就是化学品的内部结构也是一个有趣的组织，大家可以看到万事万物似乎都是有联系的，所以这个模型就是从大自然当中发展出来的，这是一种全新的概念，它是一种超分子的化学，可以有一些新的形状为我们所用。对于这种模型的研究也是非常有意思的，实际上这是一种非传统的工程方面的一些挑战。首先要用这样一种反应式的研究模式，我们发现这个系统也有一些弱点，于是我们就有了第二种模式，就跟电网一样，电网相互之间会有影响，这比第一种方式的强度要好得多，但是它的一个弱点就是不能模拟特定的环境。第三种模式就是让不同的成分之间进行相应的互动，它也能够非常好地模拟这样一种社会环境、社会体系。对一些不同的设计，这些分析也是非常重要的，有一种新的体系进入了化学工程。

自我改善的这样一个概念，不断地用来帮助我们解决这种生产里面碰到的非常复杂的矛盾，这些复合性、复杂的物质往往是最具有挑战性的，这也就是我们建立了这么多的模型、进行相应演练的一个原因所在，而且这些对于机器、对于产品、对于人和机制之间的互动都提出了更高的要求。我们不同学科之间要时时刻刻地相互关注，关注在我们临近学科取得一些什么样的进展，因为这些进展对学科发展是非常重要的。

**K.V.Raghavan**　印度工程院院士、印度化学工程师学会会员、A. P. 科学院院士、UGC杰出学者。获得奥斯马尼亚大学学士学位，马德拉斯印度理工学院(IIT)的硕士和博士学位。1964年，入职CSIR并工作于3个国家实验室；1994年，被任命为金奈CLRI研究中心主任；1996年，在海得拉巴担任印度化学技术学院主任；2004年5月，被任命为DRDO、国防部和印度政府的招募和评估中心主席；2008年10月，被授予印度化学技术研究所杰出教授。他的研究小组主要从事绿色过程的反应工程研究、化工过程开发与设计、反应工程、仿真建模和化学危害分析。他的主要研究贡献包括固定床反应器中的复杂反应模拟、多相反应系统的流体力学、清洁处理的环境催化、大分子沸石催化、热化学、电荷转移聚合动力学和化学事故的建模动力学。目前的研究领域为水煤气转换反应的过程强化、催化 $CO_2$ 分解、分析 $CO_2$ 捕获技术和离子液体反应的表征等。

# 能源与矿产资源的发展和利用：过去、现在及未来

Z. Xu, M. Wyman, J. Masliyah, K. Cadien

艾伯塔大学化工与材料工程系，埃德蒙顿，艾伯塔，加拿大

**摘要**：文明的发展始终依赖于能源和矿产资源。在百万年前的石器时代，当人类利用锋利的石块来防御和狩猎食物时，人类文明的发展就已经与通过萃取或制造的自然资源加工和利用的发展紧密相连。接着，青铜时代和铁器时代的持续发展，最终促进了玻璃、陶瓷、现代金属，如钢/不锈钢和合金/超合金、聚合物、金刚石、复合材料以及硅的发展，这些材料促进了当代电信的发展，把我们带到了信息时代。通过纳米技术探索自然，生命科学正取得快速发展。我们所有的进步都依赖于自然矿物资源和化石燃料的使用。世界人口的持续增加和对更高品质生活的要求对不可再生的矿产资源和能源资源的利用产生了巨大的压力。金属回收利用虽然能部分缓解对来自自然资源的金属的需求，但依然不能满足人口增长的需求和改善生活方式的趋势。石化资源不可再生、不可回收的性质使其面临更严峻的挑战。因此，我们在寻找替代化石燃料的同时，被迫开发越来越低品位的矿产资源。

在本文中，我们将对矿产资源的发展和利用在人类文明的进步和能源的广泛利用中起到的关键作用做一个历史性的总结。随着易加工自然资源的迅速枯竭，低品位、复杂矿产资源的加工成为必然，这将需要更高的能源消耗，并对我们的环境产生更严重的威胁。人均能源利用率日益提高，再加上不断增加的人口，要求人类找寻替代能源，特别是太阳能能源，这又要求革命性材料的发展。同时，非常规能源正变得越来越重要，以弥补现有的化石燃料为基础的能源与如太阳能等可再生能源的有效获取之间的差距。例如，在可再生资源在经济上可行及大规模需求具有竞争力之前，油砂作为可替代化石能源满足能源需求。

# 一、引言

材料一直是人类发展和进步的关键因素，追溯至史前，其作为制造工具的证据早有文字记载。今天，有一句话是："没有材料，就没有工程"，所言不虚。有证据表明，人类长久以来一直使用木材作为狩猎长矛和替代石器工具。图 1 显示了新材料被广泛采用的近似时间，这与人类文明的重大进步相一致。值得注意的是，随着新材料的广泛使用，旧材料常常以更加有效和创新的方式继续使用。例如，今天我们还用木头做建筑结构和房屋建造，并制作家具。

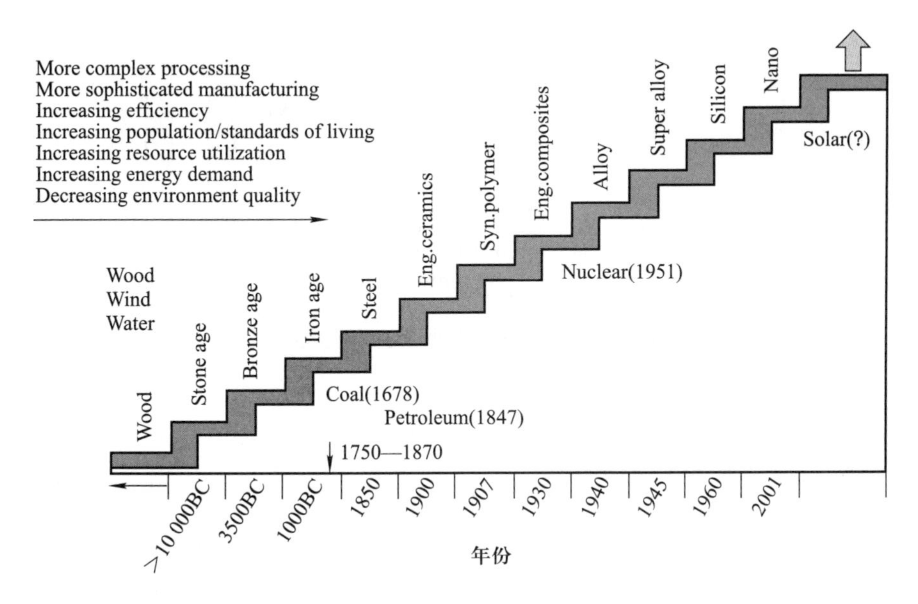


图 1　新材料被广泛采用的近似时间

# 二、过去

12 000 年前，当农业开始的时候，木材和石头等材料作为工具和武器已经使用了数百万年。用石头和木头制造工具和武器需要技术，但该技术是相当简单的。从青铜时代开始（公元前 3500 年），加工过程需要更复杂的技术。制造青铜需要铜冶炼技术及铜锡合金化技术的发明和发展。当然，世界各地的不同社会在不同时期进入了青铜时代，这取决于矿石的可用性，或者说拥有青铜技术的地方的青铜贸易的能力。在世界部分地区，铁器时代（约始于公元前 1000 年）与青铜时代重叠，但一般认为铁器时代在青铜时代之后，同时，在青铜时代，铁的使用是很普遍的。

铁矿石的加工过程需要 1200℃ 以上的温度以及真空炉的发展，早期的铁器由流星铁锻造。青铜和铁器时代的主要驱动力之一是品质更好的剑和矛的发展，也就是说,用于狩猎和保护人口增长的战争武器的发展,如图 2 所示。这两个时代都持续了 2500 年以上,但 18 世纪早期利用焦炭生产铸铁技术的发现和 19 世纪石油的发现,加速了创新步伐。

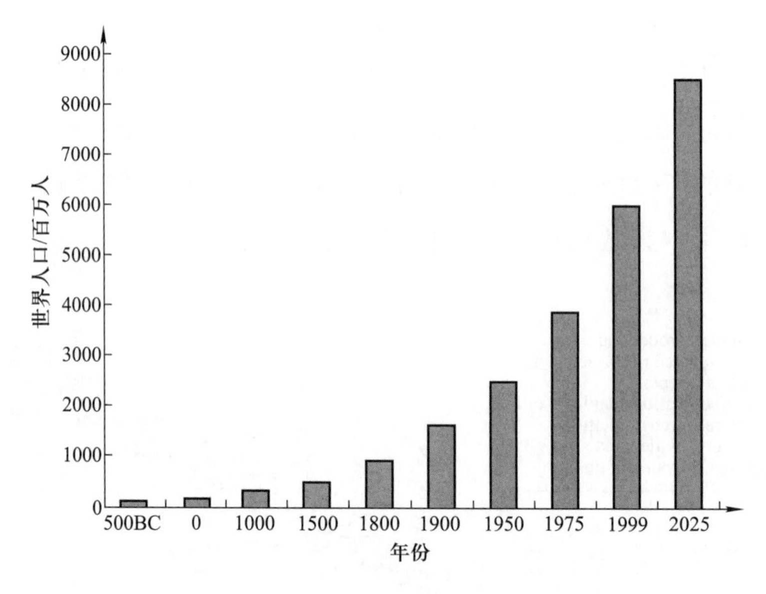


**图 2　截至 2025 年世界人口的增长**

技术创新的加速始于 18 世纪末期,可以发现,材料之所以在当时得以快速发展,是因为当时的人们需要越来越多的能量。煤炭和石油等廉价和高密度的能源加速了材料的发展,这与人类对材料如何及为什么有特殊功能的兴趣息息相关。显微镜的发明导致了金属微观结构的发现,这反过来又加速了铸铁的改善和对铁的作用的认识。采用氧消除碳的钢的发明是一个重大的突破,其推动了 19 世纪中期的工业革命。材料的发展使蒸汽的利用成为可能,随着铁路、发动机及蒸汽动力工厂的发展,其开启了一场在基础设施方面的革命。蒸汽的产生需要能量来源(热),这是通过燃烧木材或煤来实现的,从而导致了污染的增加。19 世纪末大量石油的发现促进了能源的发展,而石油能源密度是木材或煤炭的两倍,如表 1 所示。

**表 1　不同燃料的特定能量**

| 燃料 | 特定能量/(MJ/kg) |
| --- | --- |
| 柴油/燃料油 | 48 |
| 丙烷/丁烷 | 46.4 |
| 汽油(石油) | 44.4 |
| 煤 | 24 |
| 木材 | 16.2 |

由于石油是一种液体，它比煤更容易提取，并且比煤的当量体积要小得多。此外，液体燃料更容易储存和运输，并可以在高强度能量的系统中使用。石油的发现和利用促进了内燃机的发展和创新，其反过来推动了汽车、公共汽车、飞机以及柴油电动火车的发展。内燃机也促进了金属制备技术的不断改进，以及新的合金和工程陶瓷的发展。石油也加速了聚合物的发明和发展。

近 60 年来，人类文明一直被半导体技术的发明所推动，特别是硅晶体管和集成电路的发明，我们称这个时代为硅时代。如今，集成电路在各个领域无处不在，使地球上所有人的连接得以实现。半导体销售总额为 3300 亿美元/年，使全球电子市场超过 3 万亿美元。前四个国家的电子市场如图 3 所示。这个市场的发展不仅基于晶体管和集成电路的发明，也基于贝尔实验室 Pfann 发现的半导体纯化技术的突破性的进展。

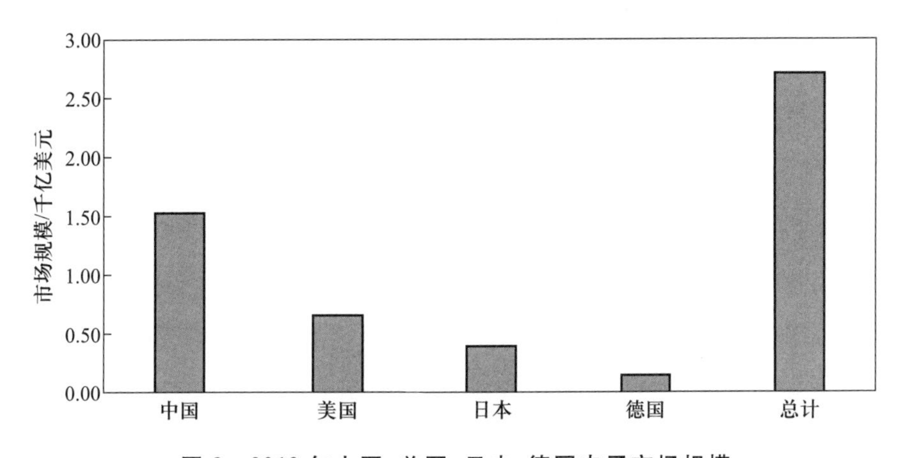


**图 3　2012 年中国、美国、日本、德国电子市场规模**

晶体管的发明恰逢人类对人类和工业对环境影响的逐渐深刻认识之时。实际上，Rachel Carson 的有关环境的开创性的书——《寂静的春天》，在 1962 出版，两年后，集成电路发明了。自此以后，对环境的忧虑吸引了人们新的兴趣，即更有效的制造方法，以及新材料的应用，以解决环境问题，如能源消耗和全球变暖。

## 三、目前

走向未来，人类将继续为更好的生活而努力，同时人口将继续以图 2 所示的指数速度增长。人类对来自自然资源材料的需求会相应增加，而非可再生自然资源却日益枯竭。为了满足日益增长的物质需求，人类不得不利用低品位、更具挑战性的自然资源，这需要更高的能源强度来发展。虽然有效地回收材料，在一定程度上将缓解材料的需求量不断增加的压力，但是材料的回收也需要使用不可再生能源资源。尽管我们尽最大努力开发和实施可再生能源，如太阳能、风能、水能、地热能和生物质能等，截至2040年，我们将继续看到人类对不可再生的化石燃料的逐渐增加的需求，如图4所示。这是由于人口(图2)的预期增长以及中国和印度等人口大国人们的生活水平不断提高。

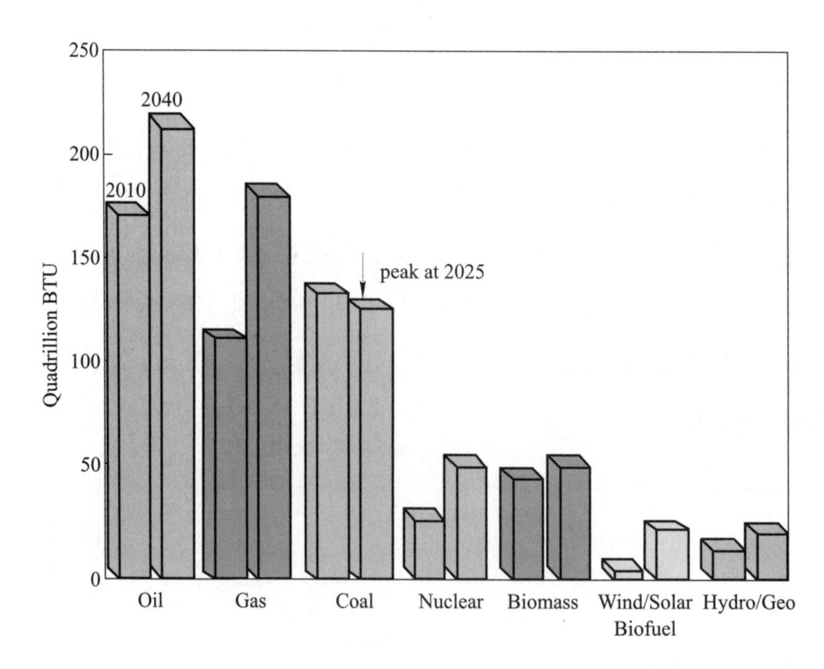


**图4　能源展望:展望2040(埃克森美孚,2014)**

在可预见的未来，非可再生能源需求的增加会促使石油行业寻找非常规石油资源。为开发加拿大丰富的油砂资源和压裂页岩去实现美国遥不可及的石油储备而导致的技术的快速发展，便是这一战略方向很好的例子。不幸的是，利用这些类型的能源资源，需要更高的能源强度的操作条件，或者将造成更严重的空气和水污染的环境问题，同时将会增加淡水供应的压力，威胁自然生态系统。

以加拿大利用微萃取开采油砂为例，生产市场化的燃料需要大约20%的能耗，同时消耗约三倍体积的水。如此高的能耗与由油砂生产石油的复杂过程相关，如图5所示。按照图5(a)的方法，开采的油砂矿石经破碎后，加入化学添加剂和热水制成浆液。因为大量高黏度的沥青在室温下不流动，所以热水是必需的。浆液通过浆体管道输送，在输送管道中，油砂经过"消化"，从油砂中释放出沥青，这被认为是一个能够使油砂工业化的创新技术。浆体输送过程中，释放的沥青液滴附着在夹带的空气中，形成密度比水的密度低得多的加气沥青。经过几公里的水汽输送，经过调节的浆料进入一个固定的分离器，被称为初级分离单元，加气沥青浮到泥浆顶部，该沥青泡沫在去除空气后，通常含有质量分数为60%的沥青、30%的水和10%的固体物质。为了进一步去除固体和水，有机溶剂，通常是沥青升级的副产品被添加到沥青泡沫中去溶解沥青，通过增加有机相和水相的密度之差，达到有效相分离，接着，在斜板、水力旋流器或离心机等的重力作用下进一步分离，统称为泡沫处理。有机相被送到有机回收单元进行回收再利用，同时，生产得到水和固体质量分数低于 1% 的沥青，其可以再进行下一步升级，如图 5(b)所示。初级分离单元和泡沫分离单元的水相被排放到大容量的尾矿池中，在此，固体和水分离得以循环利用。这种生产工艺可以得到 90% 的沥青收率，即利用 2 t 含 9%(质量分数)沥青的油砂可以生产 1 桶沥青。

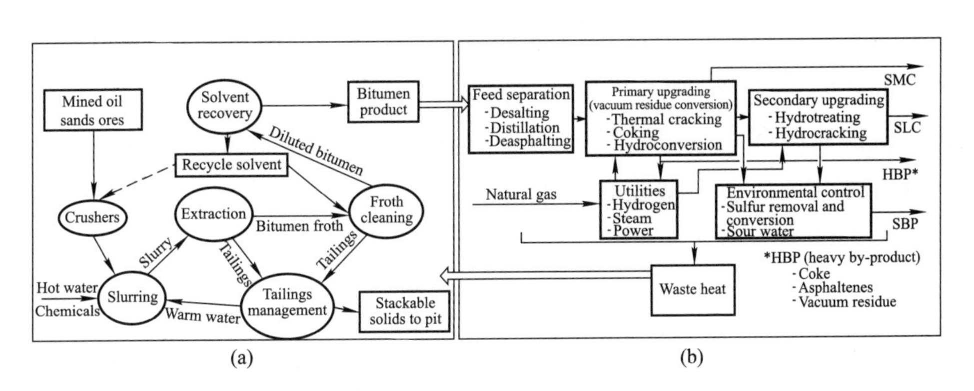


**图 5　加拿大油砂生产石油工艺图**

(a)微萃取;(b)沥青升级

在加氢裂化生产合成原油之前，通过焦化或热解对提取的沥青进行升级。焦化和热裂解主要是降低分子的大小，而加氢处理主要是从原油去除硫和氮以生产适用于精炼的产品。在升级过程中，需要大量的氢气，其由天然气中得到，而废物产生的热量可用于提取过程。

显而易见，油砂生产原油需要付出巨大的努力，不仅需要更多的能量和更复杂的设备，还会产生废尾矿和大量副产品，以及温室气体排放。相似的挑战也存在于页岩气的压裂中。不幸的是，在这一技术可以为人类提供足够的清洁能源之前，为了填补现在和未来之间的差距，我们不得不使用如图 4 所示的化石燃料。当我们在发展利用可再生资源的新材料和新技术时，我们致力于发展技术，以使得化石燃料的生产和使用对环境所造成的影响最小化。方法之一是降低萃取过程的操作温度，一个简单的能源平衡分析表明，操作温度每降低 1℃，能量将降低 10 MJ，降低的能量可生产一桶油。通过微萃取法，以每天生产 800 000 桶沥青的速率计算，如果有人能够将目前的操作温度从 45℃ 降低到 20℃，减少 5208 $m^3$ 汽油的使用量将很容易实现。这样一个能量的降低，不仅仅能够降低操作成本，也将每天减少 13 000 t $CO_2$ 气体排放。

仔细看图 5 中的操作参数，热能(热水)是用来降低沥青的黏度，使它流动，可以使其释放。降低沥青黏度的另一种方法是在泡沫处理中向沥青中加入溶剂。如图 5(a)所示,在泡沫处理中加入溶剂。一个想法是加入部分溶剂[图 5(a)中虚线箭头所示],使得沥青黏度降到和加入热水一样。这种替代方法将会消除热水的使用,称为水-非水混合提取的新工艺,示意图虚线箭头见图 5(a)。该工艺已在实验室规模被证明是可行的,这将提供一个不需要其他化学添加剂的生产工艺。通过实施沥青改质中石油焦的气化,可以实现类似的节能。这样的能源集成系统,不仅减少天然气的使用,同时产生高浓度的 $CO_2$ 气流利于收集或增强石油的回收。这个油砂操作实例在如何减少利用化石燃料带来的环境影响方面给我们提供了很好的指导,以满足我们在未来几十年里的能源需求。更重要的是,如果我们要提供能源和自然资源的安全保障,无论是可再生的或不可再生的,我们不能过分强调基础研究对革命技术发展的关键性意义。

## 四、未来

回顾历史,材料的发展推动了人类时代的发展。在本节中,我们不会试图预测未来,但我们将回顾一些仍能在未来继续的趋向。在过去的 10 年中,由于对环境的担忧和对可再生替代能源以及更有效地利用资源的追求,40 多年前的新发现已经被用于技术的革新,如图 6 所示。

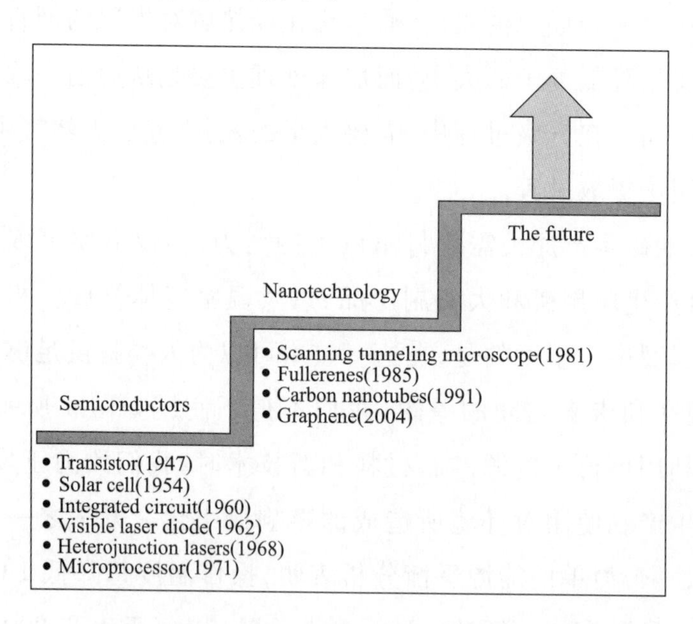


**图 6　在过去的 60 年里,创新和发现的阶梯图**

发光二极管(LED)和太阳能电池技术,分别在 1962 年和 1954 年发明。这两种技术都需要近 50 年才能对人类产生重大影响。在过去的 20 年里所做的发现,即碳纳米管和石墨烯,已经让一些人预测未来的材料领域碳时代的到来。然而,这些材料的真实效果目前仍不可知。有证据表明,随着生物技术和纳米生物学的进展,可能会生产具有突出性能的这些材料。然而,从 2014 年的观点来看,人类对更有效地利用能源、使用更高效的设备、新的可再生能源来源、更有效的内部燃烧引擎、回收更多材料的需求将继续增长,这将有可能形成一个可持续发展的未来。

## 参考文献

English Heritage. Introduction to heritage assets: Pre-industrial ironworks. https://www.english-heritage.org.uk/publications/iha-preindustrial-ironworks/preindustrialironworks.pdf.

Gosselin P, Hrudey S, Naeth M, et al.2010.Report of The Royal Society of Canada Expert Panel: Environmental and Health Impacts of Canada's Oil Sands Industry.Royal Society of Canada, Toronto.

Harjai S, Flury C, Masliyah J, et al. 2012. Robust aqueous-nonaqueous hybrid process for bitumen extraction from mineable Athabasca Oil Sands.Energy Fuels, 26: 2920-2927.

http://www.semiconductors.org/industry_statistics/global_sales_report/.

http://www.statista.com/.

Masliyah J, Czarnecki J, Xu Z.2011.Handbook on theory and practice of bitumen recovery from Athabasca Oil Sands. Vol.1: Theoretical Basis.Kingsley, Calgary.

Michael F Ashby.2011.Materials selection in mechanical design.4th Edition.Burlington M A: Elsevier.

National Geographic. The Genographic Project. The Development of Agriculture. https://genographic.nationalgeographic.com/development-of-agriculture/.

Pfann W G.1952.Principles of zone melting.Trans.American Institute of Mining and Metallurgical Engineers, 194: 747-753.

Smithsonian National Museum of Natural History.http://humanorigins.si.edu/evidence/behavior/tools.

The outlook for energy: A view to 2040 (2014, ExxonMobil). http://cdn.exxonmobil.com/~/media/Reports/Outlook%20For%20Energy/2014/2014-Outlook-for-Energy.pdf.

**Z.Xu(徐政和)** 加拿大工程院院士,加拿大艾伯塔大学教授。分别于 1982 年和 1985 年获得中南矿冶学院矿物工程专业的学士和硕士学位;1990 年,获得美国弗吉尼亚理工大学材料工程专业的博士学位。曾担任弗吉尼亚煤炭和矿产加工中心研究助理和加州大学圣塔芭芭拉分校的博士后。1992 年 9 月麦吉尔大学冶金工程专业的助理教授;1997 年 1 月阿尔伯

塔大学化学和材料工程系副教授，2000年晋升为教授。2002—2007年，担任NSERC-EPCOR-AERI工业研究主席；2007年，担任加拿大矿物加工研究首席科学家；2008年，担任NSERC-Industry Research油砂工程主席。研究领域为界面科学应用于自然资源处理和利用。已发表近300篇论文和57篇技术会议论文，获得3个美国专利和1个加拿大专利。2007年当选Tech教授和加拿大工程院院士；2010年，担任CIM研究员；2012年和2013年分别荣获“APEGA Frank Spragins奖”和“Teck环境奖”。目前，担任加拿大CIM冶金协会第三届副主席。

# 用生物技术的钥匙开启矿产资源利用的大门

邱冠周　曾伟民

中南大学资源加工与生物工程学院，湖南长沙；
生物冶金教育部重点实验室，湖南长沙

**摘要**：当前全球高品位矿产资源不断减少，矿产资源贫化现象日益严重，传统的选冶技术不能经济有效地处理低品位矿产资源，而生物湿法冶金技术在处理复杂、难处理、低品位矿产资源上显示出较大的优势。本文阐述了生物湿法冶金理论研究的三个重要突破——从宏观到微观、从定性到定量、从现象到本质，探明了生物冶金过程中的微生物种群结构、群落动态和功能活性，揭示了微生物-矿物界面的作用机制，为生物浸出机制的阐明奠定了基础，并实现了从理论到实践的成果转化，在铜、金、铀等资源领域实施了生物冶金技术的产业化，获得了较高的经济效益和社会效益，为保障国民经济的可持续发展提供了金属资源的保障。因此，全球生物冶金工作者一致倡导建立国际生物湿法冶金协会，旨在“用生物技术的钥匙开启矿产资源利用的大门”。

**关键词**：生物冶金；低品位资源；基因芯片；界面作用；工程化应用

## 一、引言

矿产资源是世界发展重要的物质基础，随着高品位资源的不断开采利用，矿产资源的贫化现象日益严重，传统采矿、选矿、冶金工艺处理这些低品位矿产资源时，效率低、流程长、生产成本高、环境污染严重，导致世界上这些低品位难处理矿产资源利用率低，难以从低品位矿石中经济有效提取 Cu、Zn、Ni、Co 等几种主要金属。

国民经济持续稳定发展，要求有色、黄金、铀等矿冶工业同步增长。但国家金属矿产资源保障程度严重不足，能源消耗不断上升，资源和能源瓶颈问题突出，严重制约国民经济和国防建设的发展。因此发展高效、环境友好的低品位矿石处理新技术势在必行。

生物湿法冶金技术是利用微生物将矿石中的有价元素选择性浸出,制备高纯金属及其材料的新技术,具有流程短、成本低、环境友好等优点,已成为世界矿物加工的前沿技术,是降低边界品位、扩大可利用资源总量的重要途径。如图1所示,从1970年开始,随着高效萃取剂的成功应用,世界上通过湿法冶炼工艺(即浸出-萃取-电积工艺)的铜产量开始不断提高。到21世纪,世界上湿法冶炼的铜产量已经超过世界总铜产量的20%,在美国、智利等国家甚至超过了30%。

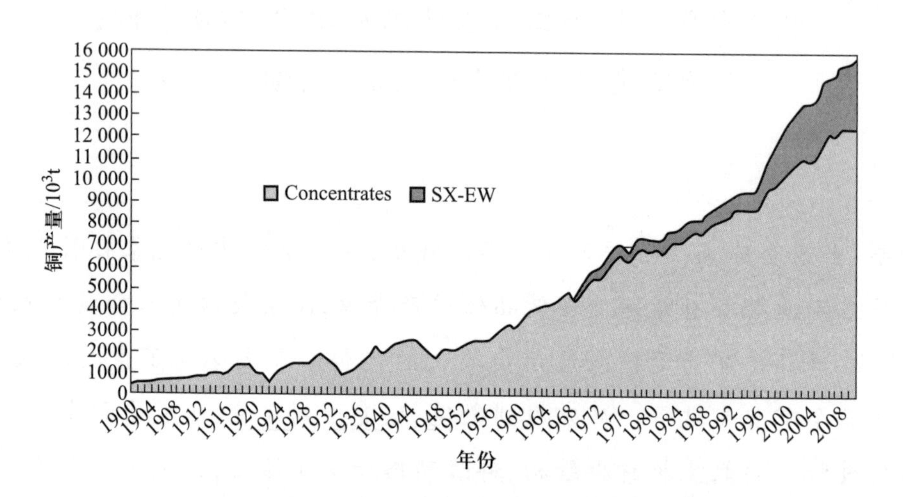


**图1 1900—2009年世界铜产量发展历程(来源:ICSG)**

深灰色区域表示生物湿法冶金产铜量

生物冶金技术是处理低品位铜矿、铀矿石和难处理含砷金矿石等矿物资源的首选工艺,该技术在国外已经实现工业化生产,在我国经过多年的研究,技术上已发展到了一个新的阶段,虽然次生铜矿、金矿、铀矿的生物浸出也逐渐进入了工业化生产,但在发展的同时也暴露出理论和工程技术上的一些问题。例如,①原料矿物的多样性、生物代谢及反应的多样性、微生物种群和功能的多样性等导致微生物浸出机制研究的滞后;②工艺参数与浸出反应过程操作条件的多样性缺乏合理性的匹配等导致工业生产中金属浸出速率和浸出率的下降。因此,国内外生物冶金领域专家学者以生物冶金技术理论、装备以及工程化条件控制等方面为研究重点,从理论到实践广泛开展研究,从而阐明生物浸出机制,研发工程化技术及装备,实现工艺条件与反应过程参数的最优化匹配,最终提高生物冶金技术的应用深度与广度。

## 二、生物冶金技术理论研究代表性成果

### (一) 从宏观到微观

在中国，长期以来生物湿法冶金研究仅停留在宏观的层次。冶金学家通常是通过不同的颜色选择效果较好的矿坑水用于浸矿(图 2)，但无法知道矿坑水中含有哪些有用的微生物以及它们是如何发挥作用的。而微生物学家只注意分离培养和显微镜观察等方法(图 3)选择优良菌种和菌株(图 4)加入浸矿体系，却不能确定微生物进入浸矿体系以后的生存情况及其功能和浸矿效果。

图 2　深红色的酸性矿坑水

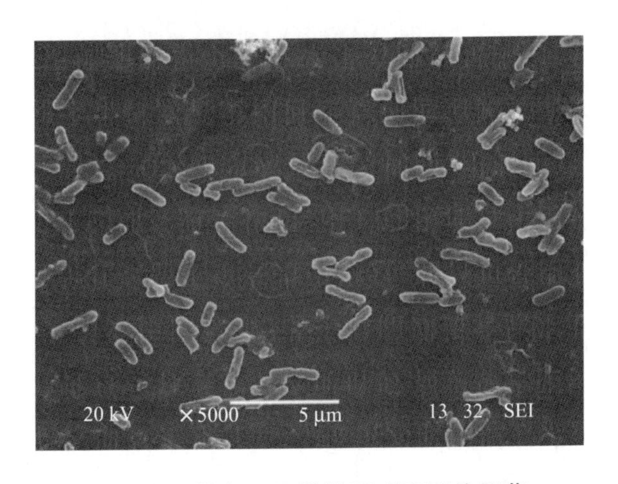


图 3　扫描电子显微镜下观察到的细菌

2004 年，中南大学参加世界上第一个嗜酸氧化亚铁硫杆菌的全基因组测试研究工作。在全基因测序获得全部 3217 个基因信息(图 5)的基础上，通过全基因组芯片和比较基因组学研究，发现了 320 个高氧化活性的基因，其中包括 135 个亚铁氧化、硫氧化以及抗性相关基因(图 6)。以此为基础制定了《嗜酸氧化亚

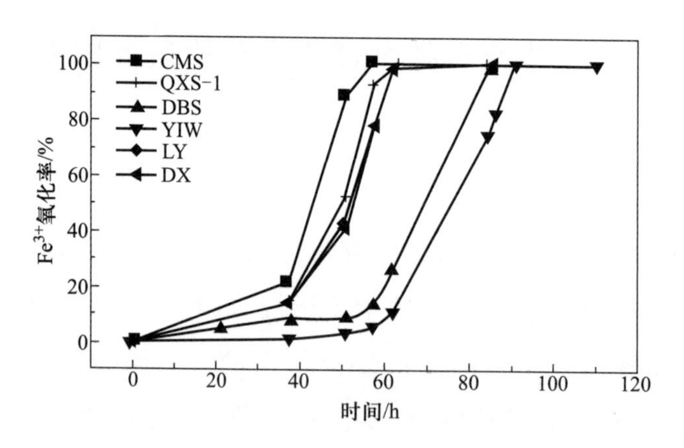


**图 4** 不同 *A.f* 菌株在氧化活性和抗性等方面的差异导致其浸矿效果不同

铁硫杆菌及其活性的基因芯片检测方法》国家标准(GB/T 20929—2007),实现了高效浸矿菌种的快速准确筛选(图 7、图 8)。嗜酸氧化亚铁硫杆菌全基因组图谱及其注解为从基因水平开展浸矿机理奠定了基础,实现了微生物浸矿行为研究由表现型向基因型转变。

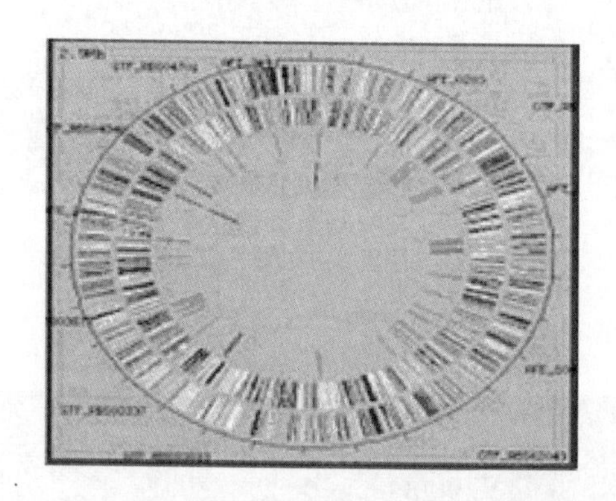

**图 5** *A.f* ATCC 23270 全基因组图谱

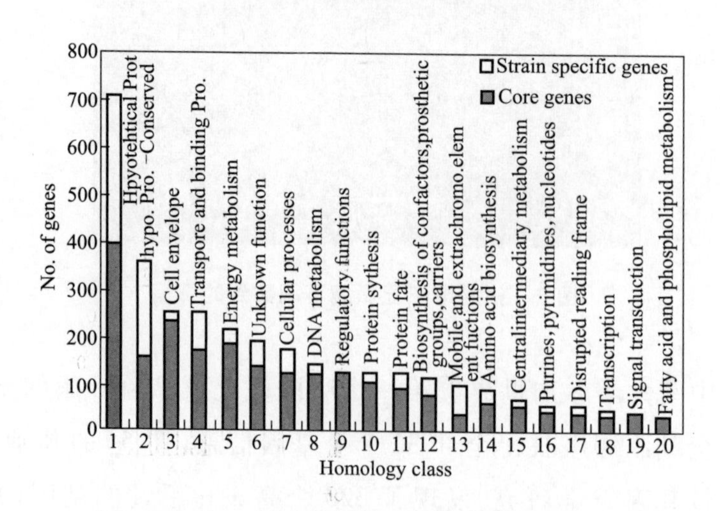


**图 6** 不同 *A.f* 菌株共有和特有基因的功能分布

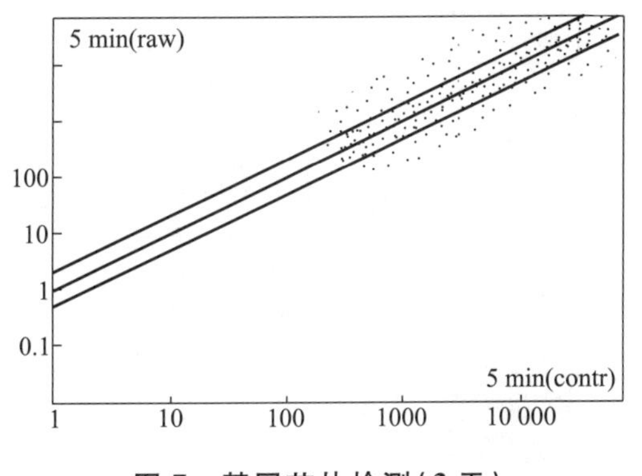


图 7　基因芯片检测(3 天)

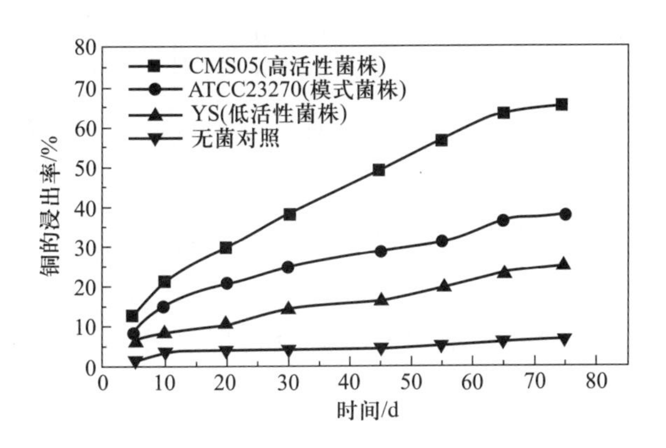


图 8　不同活性菌种的现场浸矿(75 天)

## (二) 从定性到定量

浸矿过程中工艺条件和物理化学参数的测定方法已经建立,要阐明生物浸出体系的多因素耦合机制,关键是要找到一种方法能对微生物种群和功能作用进行定量分析。但是,浸矿微生物种类多、性状与功能差异大、相互作用关系复杂、人工培养困难,以传统的分离培养和生物化学反应测定为基础的分析技术难以对浸矿微生物进行适时定量分析和实时监控。随着生物学技术的飞速发展,基因和基因组、宏基因组等技术越来越多地被应用于生物冶金领域。尤其是基因组技术的应用,使得冶金微生物的定量化有了根本性的变化。

浸矿微生物功能基因芯片和群落基因芯片技术的开发,使研究水平从单菌的单一功能提升到单菌整体功能和种群整体功能(图 9)。基于这些技术,可以定量检测出生物浸出过程中微生物种群的结构和功能的动态变化,分析生物浸出过程中,浸出参数对微生物生长及其功能的影响。

通过群落基因芯片,可以检测不同时间(图 10)、不同地点(图 11)的浸矿微生物种群结构及群落动态变化情况(图 12),通过功能基因芯片,可以分析浸矿微

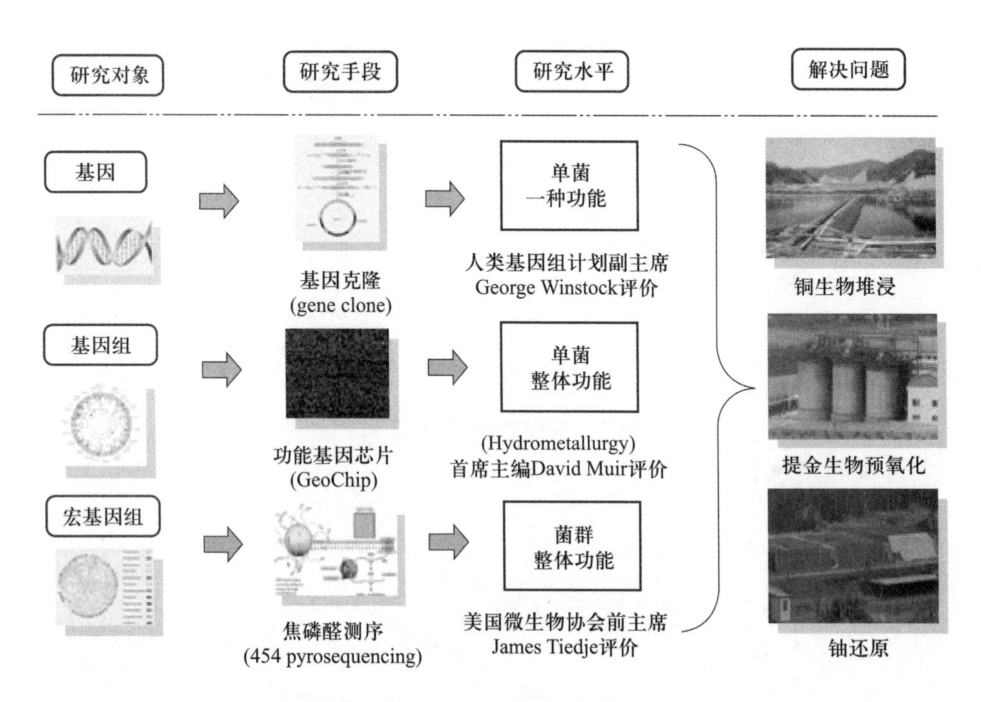


图 9　生物冶金的分子生物学技术

生物重要功能基因的表达情况(图 13),从而实现了浸矿微生物群落结构与功能活动的同步分析。

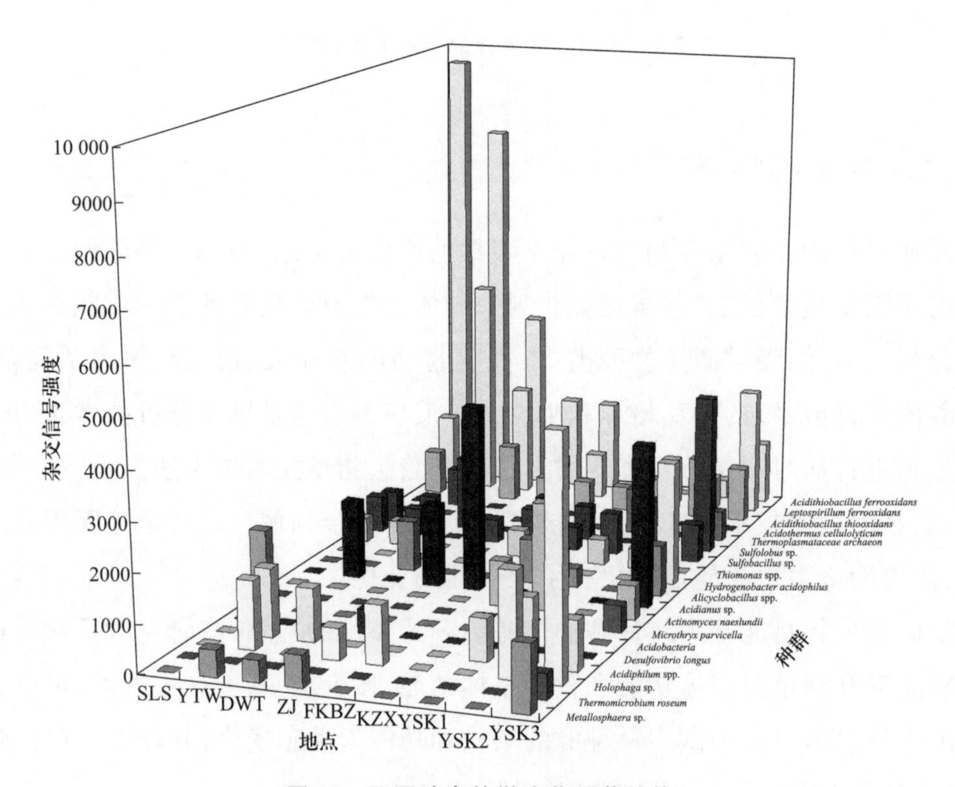


图 10　不同地点的微生物群落结构

在阐明浸出过程微生物群落结构与功能演替规律的基础上,通过不同类型

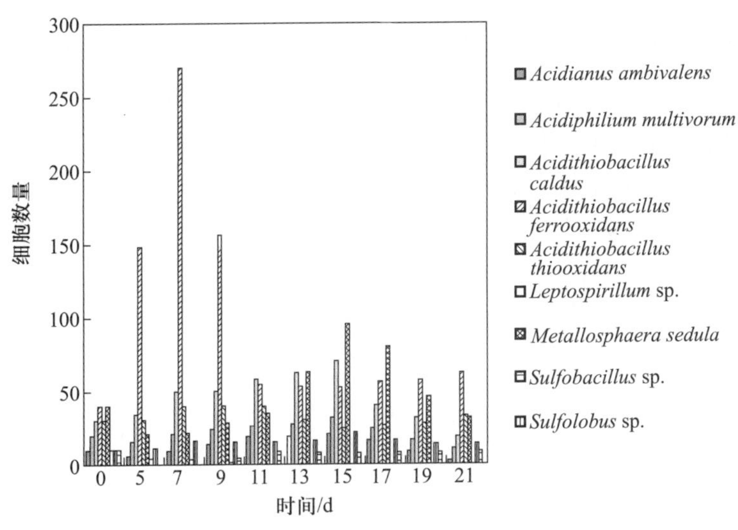


图 11　不同时间微生物种群演替规律

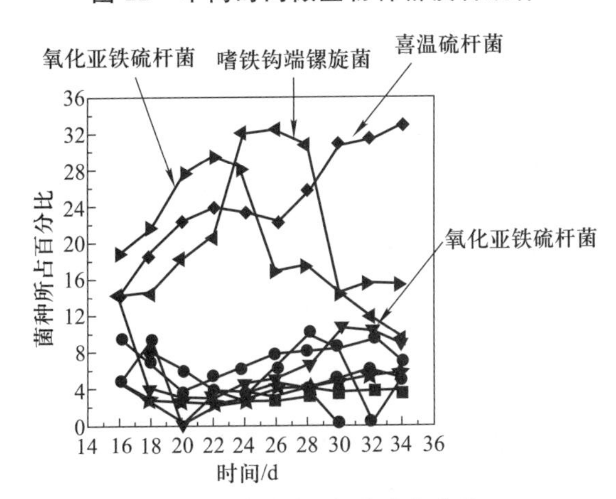


图 12　微生物种群动态变化

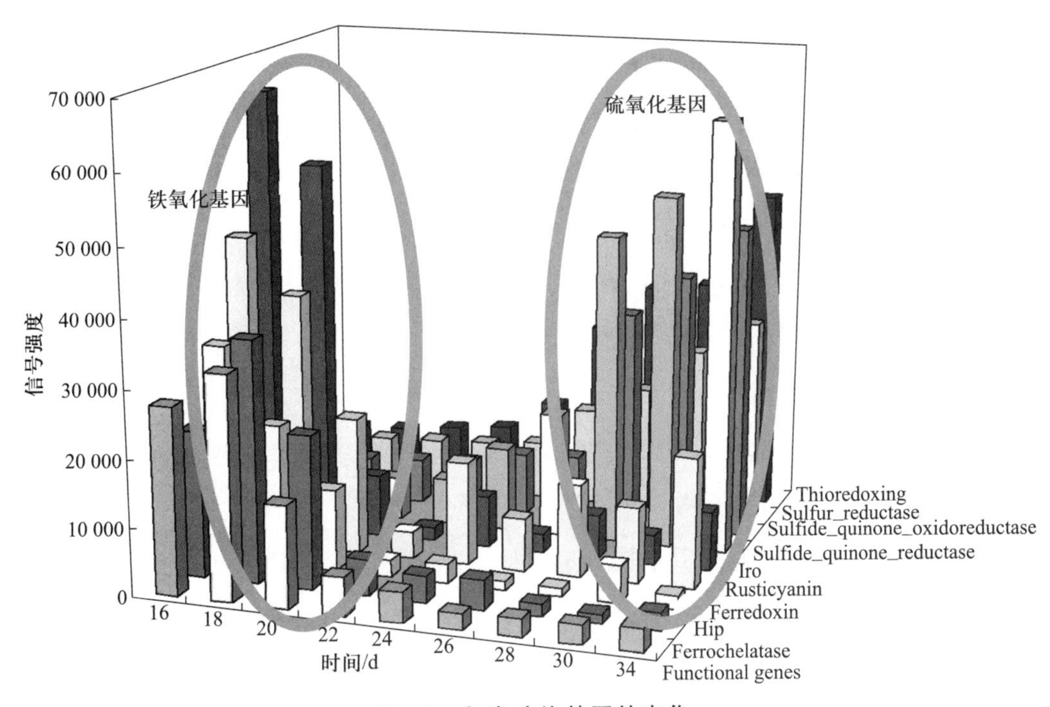


图 13　各类功能基因的变化

菌种的搭配，优化了种群组合，用于广东梅州玉水硫铜矿低品位铜矿的生物浸出，显著提高了浸出效率。如表1所示，采用优化后的种群组合Ⅳ浸出低品位黄铜矿，55天内铜的浸出率可以达到75%，相比其他组合提高了10%以上。

**表1 4个优化的种群组合及对示范工程矿样浸矿效果**

| 菌种 | *A.f* | *Lep.* | *Acidip.* | *Actino.* | Acido. | 浸出率 | 浸出时间 |
|---|---|---|---|---|---|---|---|
| 组合Ⅰ | 17.5% | 23% | 8.7% | 1.6% | 10.3% | 55.23% | 55天 |
| 组合Ⅱ | 37.5% | 22.7% | 4.5% | 9.8% | 7.1% | 57.21% | |
| 组合Ⅲ | 16% | 29% | 2% | 0% | 43% | 62.25% | |
| 组合Ⅳ | 12.1% | 16.9% | 38.7% | 1.7% | 5.8% | 75.11% | |

注：*A.f*：嗜酸氧化亚铁硫杆菌；*Lep.*：钩端螺旋菌；*Acidip.*：异养嗜酸菌；*Actino.*：放线菌；Acido.：酸杆菌纲。

## （三）从现象到本质

矿物-溶液界面是生物浸出反应的主要场所，微生物作用下矿物氧化溶解过程及反应控制机理是生物冶金研究的难点。由于缺乏相应的技术和手段，关于生物浸出过程中矿物的溶解一直停留在现象阶段，即硫化铜矿在微生物的直接或间接作用下释放出金属离子，而对其中复杂的生物化学反应过程则知之甚少。

如要系统、科学地深入探讨矿物生物氧化溶解途径，需结合先进的界面分析设备及手段，检测微生物-矿物界面的形态和构造以及生物化学反应过程中中间产物的种类和含量等，从而阐明矿物溶解的本质。

澳大利亚墨尔本拥有世界上先进的第三代同步辐射加速器，通过其中的XRD、XPS以及AFM等设备，可以较为精确地分析微生物和矿物的作用过程以及各类中间产物的生成情况。

1. 阐明了浸矿微生物代谢活动和产物对界面作用的影响

浸矿微生物吸附到矿物表面后，能迅速产生胞外多聚物，首先包裹微生物自身，然后向矿物表面蔓延，最终形成一层生物膜（图14）。一般而言，吸附微生物胞外多聚物的主要成分为多糖和脂类。以亚铁为能源生长的微生物胞外多聚物中含有糖醛酸，而在以硫为能源生长的微生物胞外多聚物中很难找到糖醛酸，这可能与胞外多聚物中的糖醛酸具有富集铁离子的功能有关。

胞外多聚物的量一般在生物浸出前期产生较快，在浸出中期达到最大值，此时伴随着金属浸出率的显著提升。然而胞外多聚物一旦产生，在生物浸出环境中很难降解或消除，因此在生物浸出后期将有可能介导黄钾铁矾等沉淀到矿物表面，加剧矿物的钝化，抑制铜的持续浸出。

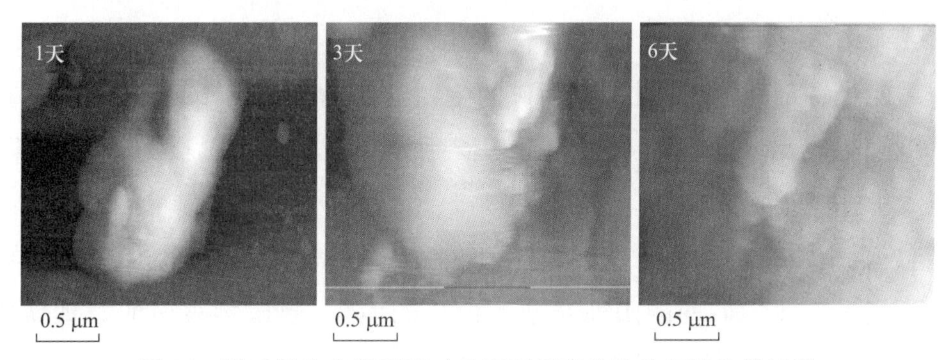


**图 14　浸矿微生物吸附到矿物表面后产生胞外多聚物的过程**

2. 证实了胞外多聚物对铁的富集作用

生物冶金研究几十年来，提出了若干生物浸出机制，包括直接作用、间接作用、接触作用、直接-间接接触作用等[17-18]。2003 年，Crundwell 通过总结前人的浸出理论，建立了一种新的生物浸出模型：他认为生物浸出应该是间接浸出、直接-间接接触浸出的合作作用机制[19]。虽然迄今为止，Crundwell 理论在生物冶金领域获得了较多人的认可，但该理论的核心机制"直接-间接接触浸出作用过程中胞外多聚物对三价铁离子的富集作用"一直处于猜想阶段，从未经过实验证明。

通过采用超声、振荡、加热等方法成功从矿物表面提取了大量富集的铁离子，包括亚铁离子和三价铁离子，另外，通过 X 射线衍射（XRD）、SEM-EDX 等分析（图 15），进一步证实了这些铁离子来自于矿物与微生物的界面上，从而首次证明了生物浸出过程中铁离子在矿物表面富集的论断，证实了生物浸出间接接触作用机制的科学合理性。

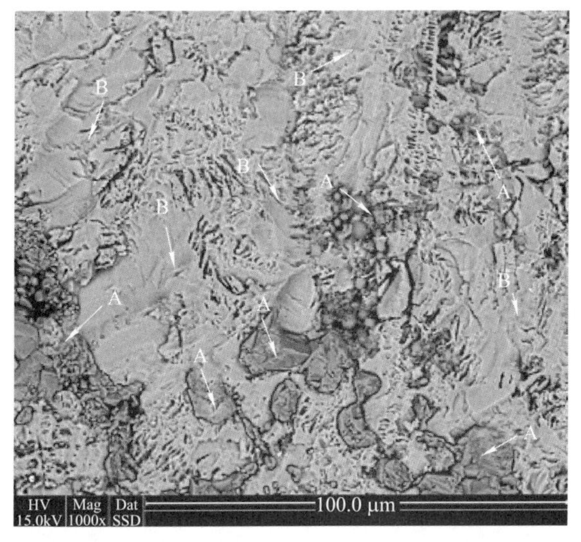


**图 15　黄铜矿纯矿矿片在生物浸出 24 天后的 SEM 图片和 EDX 结果分析**

A. 铁氧化物；B. 黄铜矿。黄铜矿矿片在生物浸出后用稀硫酸（pH 2.0）稍微清洗表面，置于空气中干燥 2 h，此后用于 SEM-EDX 分析

3. 探讨了黄铜矿的氧化溶解过程

采用同步辐射 XRD 探讨了黄铜矿的生物溶解途径(图 16)。结合文献报道,发现黄铜矿的氧化溶解过程主要分为三步:脱铁阶段、脱铜阶段和单质硫氧化阶段[20-24]。脱铁步骤中黄铜矿变成 $Cu_{1-x}Fe_{1-y}S_{2-z}(y>x)$ 是黄铜矿浸出前期的限速步骤,随后会生成多种中间产物,如 $Cu_8S_5$、$Cu_7S_4$、$Cu_{39}S_{28}$、CuS、$Cu_2S$ 等。单质硫的氧化是黄铜矿浸出后期的限速步骤。但总的而言,当矿物表面的黄钾铁矾、单质硫等日益积累,并结合胞外多聚物形成钝化膜时,黄铜矿的生物浸出将大大降低,因此此时的解钝化步骤将成为整个黄铜矿生物浸出的限速步骤。

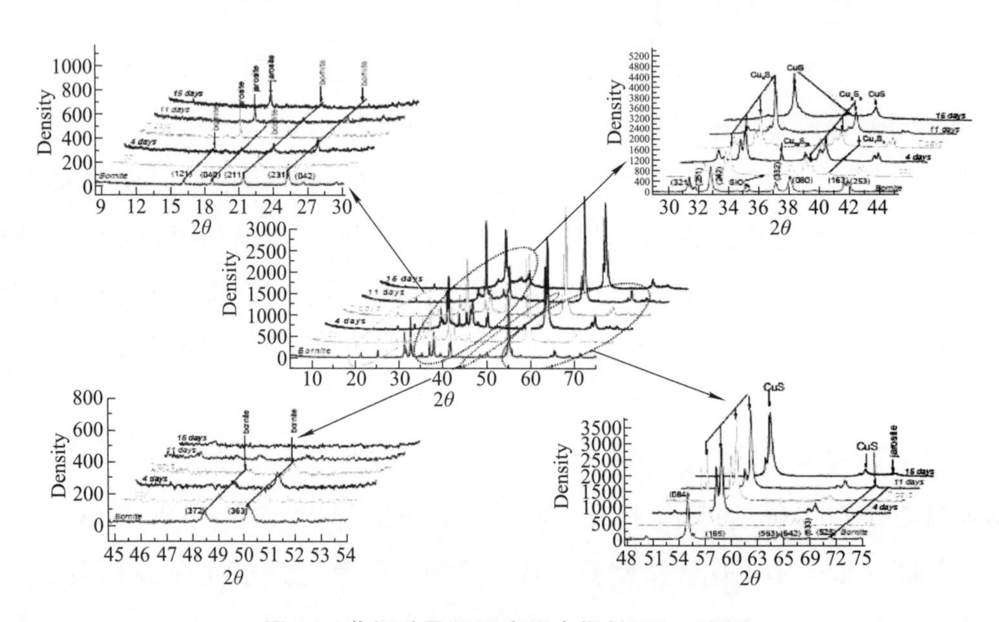


图 16 黄铜矿界面反应同步辐射 XRD 分析

## 三、生物冶金技术产业化实施现状

### (一) 铜资源的生物冶金工业应用

1. 江西德兴铜矿

中国江西德兴铜矿的生物堆浸工艺是生物冶金技术成功应用的代表之一(图 17)。江西铜业公司德兴铜矿在几十年的露采过程中,剥离了大量含铜 0.05%~0.25%的废矿石,其总量超过了 3.5 亿吨,含铜总量约为 60 万吨。由于该废矿石中的铜主要以原生硫化铜形式存在,采用传统的生物冶金方法很难获得较高的浸出率。

为了有效回收这部分铜资源,开展“德兴低品位硫化矿人工细菌浸出及萃取第三相防治方法研究”与“德兴铜矿低品位细菌浸出菌种改良和催化机理及其工

图 17　江西德兴铜矿微生物冶金工业应用

业化应用研究”。通过中南大学先进的菌种定量化技术，采用基因芯片快速筛选的方法获得了氧化活性高、抗有毒离子能力强的高效菌种，对 2000 t 电解铜的生物浸出萃取电积工厂进行了技术改造，显著提高了铜的浸出率和浸出速率。

2. 福建紫金山铜矿

紫金山铜矿是生物冶金技术在中国成功应用的典型代表。紫金山铜矿位于福建省上杭县，该地已探明硫化铜矿储量为 2.4 亿吨，铜的平均品位为 0.063%。矿石中铜矿物以次生硫化矿为主，主要包括辉铜矿、铜蓝和硫砷铜矿等。由于铜品位非常低，传统的浮选和熔炼工艺不能经济有效地处理这种矿石，而生物湿法冶金工艺是一条可行之路。通过堆浸工业化试验研究（图 18），在紫金山铜矿建立了年产 1 万吨铜的生物堆浸厂，矿石铜品位下降到 0.40%，紫金山铜矿铜金属储量由 278.41 万吨扩增到 307.46 万吨，铜浸出率为 80.11%，堆浸周期缩短到

图 18　紫金山铜矿生物堆浸工业化应用

185 天，高纯阴极铜产品直接加工成本为每吨铜 12 812.00 元，与传统的浮选-火法冶炼过程相比，生产成本低，企业经济效益、社会效益显著（表 2）。

**表 2　生物提铜工艺与常规闪速炉炼铜对比（吨铜）**

| 冶炼方法 | 资源/t | 能耗 | | 水 | 温室效应 | 酸化效应 |
|---|---|---|---|---|---|---|
| | | 电耗/(kW·h) | 标煤/kg | 新水/$m^3$ | $CO_2$/kg | $SO_2$/kg |
| 生物提铜 | 307.46 | 3915.14 | 1402.22 | 21.76 | 4090.57 | 11.93 |
| 浮选-火法炼铜 | 278.41 | 8706.90 | 3656.24 | 168.09 | 10 909.29 | 79.04 |
| 相对量 | 110.43% | 44.97% | 38.35% | 12.85% | 37.50% | 15.09% |

3. 赞比亚谦比希铜矿

生物冶金研究成果不仅在中国取得了较好的效果，还进一步推广到了国外。2010 年赞比亚矿业部与中南大学签订战略框架合作协议，并建立了“中国有色集团-中南大学赞比亚生物冶金技术产业化示范基地”，采用生物冶金技术处理该国大量的表外矿及尾矿资源。

2011 年 3 月，赞比亚谦比希湿法冶炼公司同中南大学合作开展低品位铜废矿石的生物堆浸工业研究。通过筛选、富集及驯化当地浸矿微生物，使微生物较快适应了堆场环境，并保持良好的生长状况。随后进行微生物的逐级扩大培养，培养规模从 5 L、50 L、1 $m^3$、20 $m^3$ 提高到 150 $m^3$（图 19）。将培养后的菌种接种到矿堆中，使整个堆浸体系细胞浓度显著提高，从而缩短堆浸时间，提高铜浸出速率和浸出率（表 3）。

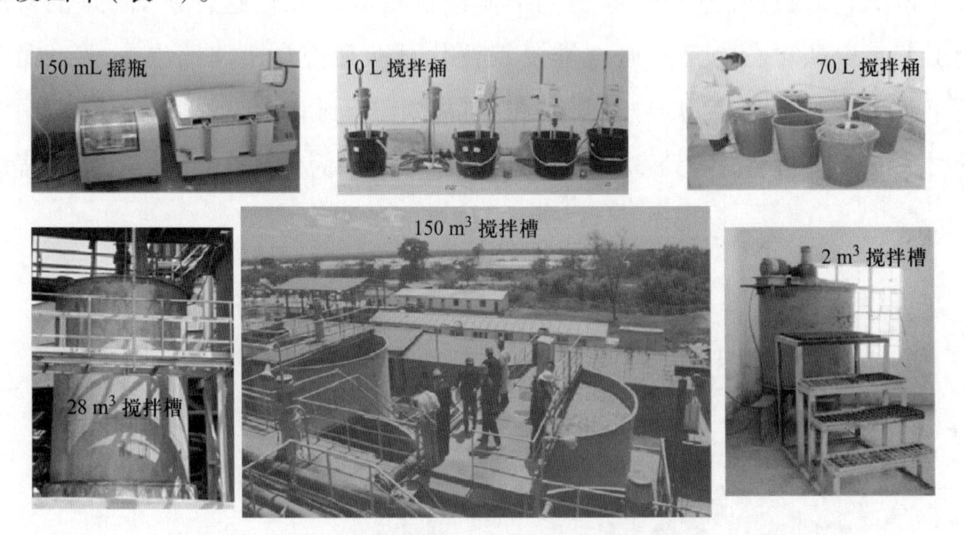


**图 19　赞比亚谦比希铜矿浸矿微生物的扩大培养流程**

在60万吨低品位铜矿石的生物堆浸工业生产中,2个月内铜的浸出率达到50%,浸出液通过后续的萃取-电积工艺,获得纯度为99.99%的阴极铜。在不改变原来工程条件的基础上,现有的生物冶金技术使谦比希湿法炼铜厂堆浸铜产量提高了20%,酸耗降低35%以上(表3),大量以前不能回收的铜资源得到了有效利用。

**表3 生物冶金和原有酸浸工艺流程参数及技术指标对比**

| 工艺参数 | 酸浸工艺指标 | 生物冶金流程指标 |
|---|---|---|
| 矿石品位*/% | ≥1.2% | ≥0.3%** |
| 浸出时间 | 4~8 h | 60 d |
| 硫酸浓度/(g/L) | 30~40 | 10~15 |
| 采用浸矿微生物 | — | 合适的菌种组合 |
| 浸出液铜浓度/(g/L) | 4~8 | 3~5 |
| 总铜浸出率/% | 65%~70% | 85%~90% |
| 吨铜消耗硫酸 | 4.8 t | 2.2 t |
| 浸出液pH值 | 1.4~1.6 | 1.8~2.1 |

* 谦比希湿法公司的生产实际;** 谦比希湿法的堆浸铜的品位为0.3%~0.5%。

## (二)金矿生物预处理工艺

中国黄金集团公司以辽宁天利金业公司含砷难浸金精矿为主攻目标,围绕着生物氧化-氰化提金技术路线,对生物氧化提金技术进行了自主创新,形成了完整的、具有自主知识产权的CCGRI生物氧化提金技术,并实施了推广与示范。经过对浸矿菌长期的定向培养、驯化,获得了与国外BIOX工艺和BACOX工艺不同的浸矿工程菌(命名为HY-系列菌)。以该菌种为核心进行相关工艺的技术开发,形成了CCGRI技术系列。建设了中国黄金行业的第一座高技术产业化示范工程;从2003年7月投产至今,已稳定运行了8年。在生产上,技术指标越来越先进;在研究上,不断有新的发现和突破。随着生物预处理技术的发展,中国黄金产量连续四年成为世界第一(图20)[25]。

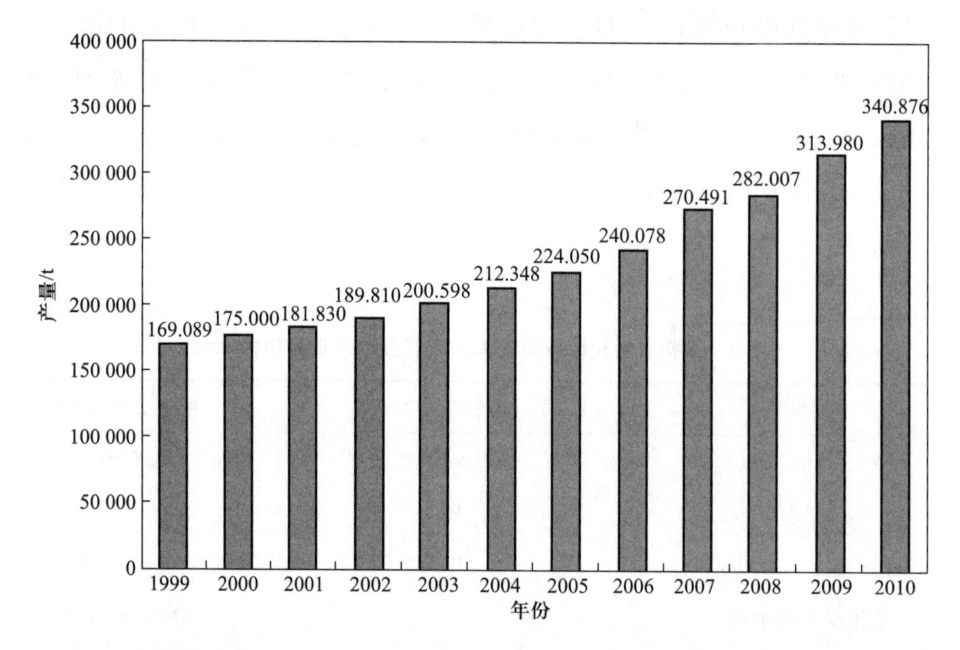


**图 20　随着生物预处理技术的发展，中国黄金产量连续四年成为世界第一**

## （三）铀矿生物提取工艺

为了满足核电发展对天然铀矿产不断增长的需求，中国天然铀生产已转向难处理铀矿石、低品位铀矿石以及与其他矿产资源伴生铀的加工。铀矿酸浸时，必须先将低价态的三价铀离子用三价铁离子氧化成高价态的五价铀。生物浸出技术是在此过程中催化氧化铁离子使其成为三价状态，从而保持整个浸出反应的完成。

铀矿生物提取技术将成为铀加工利用的关键技术。利用研发的优良菌种组合和优化工艺对江西抚州 721 矿进行堆浸工业应用，97 天铀的浸出率达到 96. 82%（图 21）[26]。生物冶金技术的进一步推广应用可使中国目前大量闲置或废弃的硫化物包裹类铀矿资源得到有效利用，并有望使中国铀矿资源开采品位从

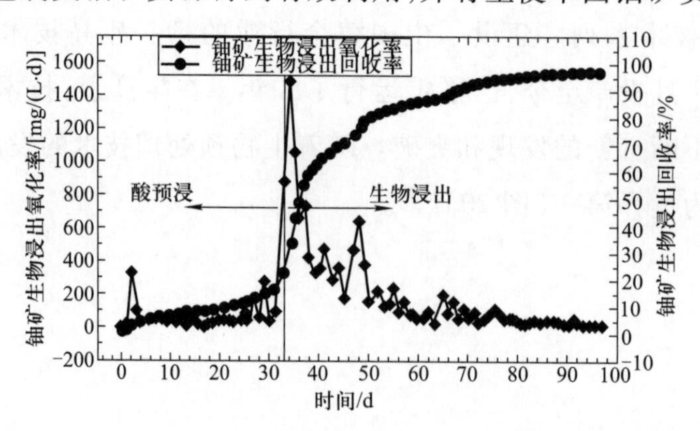


**图 21　铀矿生物浸出氧化率与回收率**

目前的千分之二降低到万分之五，从而使铀矿可经济开采储量大大增加。



## 四、展望

常规的冶金流程涉及激烈氧化反应和激烈还原反应，是一个需要大量能量消耗和产生大量含碳物质的过程，并带来严重的环境污染。而生物冶金反应主要在常温、常压进行，是温和的氧化反应和温和的还原反应。并且微生物在生长过程中利用空气中的二氧化碳为碳源合成生物体，这个过程不但不产生对环境有污染的二氧化碳，相反，它还像植物一样消耗大气中的二氧化碳，是一个减碳过程，生物冶金技术具有清洁、安全、低成本的特点：①有效扩大可用资源，可处理原矿、表外矿、废石、炉渣、燃烧灰，回收二次金属、工业电子垃圾、城市污泥等，只要有价态的变化，都可以利用微生物进行氧化和还原，从铜、金、铀的生物氧化效果来看，它既可用于硫化矿，也可用于氧化矿和复杂矿物；②广泛实用，在静态浸出（堆浸）中，可用于铜、金、铀、镍等金属的提取，在动态浸出（槽浸）中，可用于金、铜、镍、钴等的提取。

生物冶金技术在处理低品位矿产资源上具有较大的优势。在当前矿产资源贫化现象日益严重，传统选冶技术难以从低品位矿石中经济有效提取 Cu、Au、Zn、Ni、Co 等几种主要金属的情况下，要保障我国国民经济可持续发展，生物冶金技术在未来的矿物加工处理方面将占有举足轻重的地位。

因此，在中南大学主办的第 19 届国际生物湿法冶金大会上，大会主席邱冠周院士提出"用生物技术的钥匙开启矿产资源利用的大门"，并得到与会专家的一致认可，同意成立"国际生物湿法冶金学会"，并将总部设在中国湖南长沙，旨在推动全球生物湿法冶金技术的发展，解决低品位矿产资源难以利用的难题，为国民经济的可持续发展提供保障。

**邱冠周** 矿物工程学家，广东省梅州市人。1987 年 9 月毕业于中南工业大学矿物加工工程专业，获博士学位。曾任中南工业大学副校长、中南大学副校长，现任中南大学教授、博士生导师。长期致力于我国低品位、复杂难处理金属矿产资源加工利用研究，在细粒及硫化矿物浮选分离和铁矿直接还原等方面取得显著成绩，特别是在低品位硫化矿的生物冶金方面做出

突出贡献，被授予国家有突出贡献科技专家。发表了97篇科技论文和5部专著，先后获得国家技术发明奖二等奖2项，国家科技进步奖二等奖1项，国家科技进步奖一等奖1项，中国高等学校十大科技进展2项；2003年担任国家自然科学基金创新群体学术带头人，2004年、2009年连续两次担任生物冶金领域国家“973”计划项目首席科学家，担任2011年第19届国际生物冶金大会主席，并被推选为国际生物湿法冶金学会副会长。2011年当选中国工程院院士。

# 流程优化的系统方法

**Arthur Ruf**

瑞士苏黎世联邦高等工学院,苏黎世,瑞士

**摘要**:在特定的不同领域中,如化工、冶金与材料工程,大量信息是流程改善的基础,也是创新的源泉,这些是令人印象非常深刻的。未来最重要的问题是能够脱离这些巨量信息创造知识。企业为寻求竞争优势,将知识用于解决工业问题。深入了解流体动力学、传热传质动力学、传感器技术、建模和仿真化学、材料科学是解决化学、冶金和材料工程问题的先决条件。利用工业案例将是未来解决化学、冶金和材料工程问题的关键因素。

**Arthur Ruf** 瑞士工程科学院院士,瑞士工程科学院(SATW)副主席兼外事委员会主席。担任欧洲社会膜科学与技术的财务主管,瑞士技术科学院董事会成员和外交事务主管,瑞士技术科学会员,德国化学工程与生物技术协会成员。1981 年获得瑞士联邦苏黎世工学院博士学位;1981 年任瑞士联邦苏黎世工学院过程工程研究所高级助教;1984—1996 年任瑞士联邦苏黎世工学院讲师;2004 年任库尔应用科学大学讲师。主持多次国际会议,为多家公司的高级顾问。

# Part Ⅰ

# Overview of the Top-level Forum

# Overview of the Top-level Forum

June 2-3,2014,2014 International Conference on Engineering and Technology was held in Beijing. The conference was jointly organized by the United Nations Education, Scientific and Cultural Organization ( UNESCO ), the Academy of Engineering and Technology Board ( CAETS ) and the Chinese Academy of Engineering. The conference theme is " Engineering Science and Technology and The Future of Human" . This conference provided a platform for the global engineering, industry and research institutions and government participants, to share engineering frontier science and technology, travel and explore the future development direction, cope with the challenges of the future human wisdom. More than 1500 people from over 30 countries around the world took part in this conference, including the dean of the national academy of engineering, Chinese Academy of Engineering and foreign academicians, Chinese and foreign engineering technology. The meeting is divided into nine chapters, such as " Future Mechanical Engineering" ," Information Network and Social Development" ," The Future of Chemical, Metallurgy and Material Engineering" ,etc.

" The Future of Chemical, Metallurgy and Material Engineering" session was jointly hosted by the Division of Chemical, Metallurgical and Materials Engineering, Chinese Academy of Engineering and Beijing University of Chemical Technology on June 2,3.

Over 150 people from chemical industry, metallurgy and material fields attended this session,including more than 60 academicians and 80 professors. In the branch of the opening ceremony,Academician Hailing Tu pointed out that engineering and technology made great contribution to promoting economic development and improving people's living standards. This conference would mainly discuss how engineering and technology innovation make new contributions to the sustainable development of global economy and the society. He hope that this meeting will help us further strengthen the cooperation between global scientific and engineering.

Plenary lectures were chaired by Tianwei Tan,Ulrich W. Suter,Jiongtian Liu,Aibing Yu,Huibin Xu,Zhenghe Xu,Yu Zhou and Robin J. Batterham. Academician Qikun Xue of Chinese Academy of Sciences gave a speech on "Atomic Level Control of Quantum Material Growth: From Quantized Anomalous Hall Effect to High Temperature Superconductivity". Member of the National Academy of Engineering Cato T. Laurencin and Australian Academy of Sciences Academician Aibing Yu respectively gave a speech on "Regenerative Engineering,a New Field: Theory and Practice" "Simulation and Modelling of Particulate Systems". Academician Liangshi Fan (Chemical Looping Technology: Iron-Based Ohio State Processes),Academician Fu (Photocatalysis-based Novel Technologies for Clean Energy and Environment), Academician Zhang (Biomaterials for Inducing Tissue Regeneration: The New Era of Biomaterials) and Academician Qiu (Biohydrometallurgy: Biotech Key to Unlock Mineral Resources Value),Academician Helen Valerie Atkinson (Cornflour, Ketchup and Parts for Cars: A Review of Semi-Solid Processing), Academician K. V. Raghavan (The Future Product/Process Development Challenges in Chemical and Material and Allied Engineering Fields), Academician Xu (Energy and Mineral Resource Development and Utilization: Past,Present and Future),Academician Arthur Ruf (Sytems Approach for Process Excellence) clarified the new developments in chemical,metallurgy and material engineering respectively from their own views. They also discussed the innovation of engineering and the future of mankind,and how to make new contributions to the sustainable development of global economy and the society.

At the end of the session, the participating experts exchanged ideas on the advanced technology and engineering cases of the reports. Engineering technology related topics such as education,talent training,science and technology innovation were also discussed.

Academician Tu gave a summary statement and he said that "The Future of

Chemical, Metallurgy and Material Engineering" session, academic reports covered advanced materials, biotechnology, and the computer simulation system, which were all the latest progress of metallurgical, material and chemical technology. Through this actively discussion, it would make us better understand the chemical industry in the field of metallurgy and materials engineering research. It also showed us the situation including the problems and challenges and would help us to cooperation.

# Part Ⅱ

# List of Experts Attending the Forum

# List of Experts Attending the Forum

| | |
|---|---|
| **Ulrich W. Suter** | Swiss Academy of Engineering Sciences, Prof. Member of Swiss Academy of Engineering Sciences |
| **Liang-Shih Fan** | The Ohio State University, Prof. Member of the U. S. National Academy of Engineering, Foreign Member of the Chinese Academy of Engineering |
| **Arthur Ruf** | ETH Zurich, Prof. Member of Swiss Academy of Engineering Sciences |
| **Cato T. Laurencin** | University of Connecticut, Prof. Member of the U. S. National Academy of Engineering |
| **Helen Valerie Atkinson** | The University of Leicester, Prof. Fellow of the Royal Academy of Engineering |
| **K. V. Raghavan** | Indian Institute of Chemical Technology, Prof. Fellow of the Indian National Academy of Engineering |
| **Zhenghe Xu** | University of Alberta, Edmonton, Alberta, Canada, Prof. Fellow of the Canadian Academy of Engineering |
| **Aibing Yu** | Monash University, Prof. Fellow of both Australian Academy of Science and Australian Academy of Technological Sciences and Engineering |
| **Robin J. Batterham** | Rio Tinto Limited, Prof. Fellow of Australian Academy of Science, Foreign Member of the Chinese Academy of Engineering |
| **Eric Forssberg** | Lulea University of Technology, S – 97187 Lulea, Sweden, Prof. Member of Swiss Academy of Engineering Sciences, Foreign Member of the Chinese Academy of Engineering |

| | |
|---|---|
| **P. Somasundaran** | Columbia University, New York, USA, Prof. Member of the U. S. National Academy of Engineering, Foreign Member of the Chinese Academy of Engineering |
| **Hannelore Bowman** | Institute of Chemical Engineers, Prof. |
| **Yang Shen** | Columbia University, New York, USA, Prof. |
| **Qikun Xue** | Tsinghua University, Prof. Member of the Chinese Academy of Engineering |
| **Xianghong Cao** | China Petrochemical Corporation, Prof. Member of the Chinese Academy of Engineering |
| **Xiangbao Chen** | AVIC Beijing Institute of Aeronautical Materials, Prof. Member of the Chinese Academy of Engineering |
| **Yongnian Dai** | Kunming University of Science and Technology, Prof. Member of the Chinese Academy of Engineering |
| **Xianzhi Fu** | Fuzhou University, Dr. Member of the Chinese Academy of Engineering |
| **Hengzhi Fu** | Northwestern Polytechnical University, Dr. Member of the Chinese Academy of Engineering |
| **Yong Gan** | Chinese Academy of Engineering, Prof. Member of the Chinese Academy of Engineering |
| **Congjie Gao** | Zhejiang University of Technology, Prof. Member of the Chinese Academy of Engineering |
| **Jilin He** | Nerthwest Rara Metal Materoals Research Institute, Prof. Member of the Chinese Academy of Engineering |
| **Yongkang Hu** | Fushun Research Institute of Petrolesm and Potrochemicals, Sinopec, Prof. Member of the Chinese Academy of Engineering |
| **Boyun Huang** | Central South University, Prof. Member of the Chinese Academy of Engineering |

| | |
|---|---|
| **Xigao Jian** | Dalian University of Technology, Prof. Member of the Chinese Academy of Engineering |
| **Dongliang Jiang** | Shanghai Institute of Ceramics, Chinese Academy of Sciences, Prof. Member of the Chinese Academy of Engineering |
| **Wei Ke** | Institute of Metal Research CAS, Dr. Member of the Chinese Academy of Engineering |
| **Guanxing Li** | China North Nuclear Fuel Co. Ltd, Dr. Member of the Chinese Academy of Engineering |
| **Longtu Li** | Tsinghua University ,Beijing,China,Prof. Member of the Chinese Academy of Engineering |
| **Yanrong Li** | University of Electronic Science and Technology of China, Prof. Member of the Chinese Academy of Engineering |
| **Yuanyuan Li** | Jilin University, Prof. Member of the Chinese Academy of Engineering |
| **Zhongping Li** | Aerospace Research Institute of Materials & Processing, Prof. Member of the Chinese Academy of Engineering |
| **Bo Li Liu** | Beijing Normal University,Prof. Member of the Chinese Academy of Engineering |
| **Jiongtian Liu** | Zhengzhou University,Prof. Member of the Chinese Academy of Engineering |
| **Bingquan Mao** | Sinopec Beijing Research Institute of Chemical Industry, Dr. Member of the Chinese Academy of Engineering |
| **Xuhong Qian** | East China University of Science and Technology,Dr. Member of the Chinese Academy of Engineering |
| **Guanzhou Qiu** | Central South University, Prof. Member of the Chinese Academy of Engineering |

| | |
|---|---|
| **Fengting Sang** | Dalian Inst of Chemical Physics of Chinese Academy Science, Prof. Member of the Chinese Academy of Engineering |
| **Xingtian Shu** | Sinopec Research Institute of Petroleum Processing, Prof. Member of the Chinese Academy of Engineering |
| **Chuanyao Sun** | Beijing General Research Institute of Mining &Metallurgy, Prof. Member of the Chinese Academy of Engineering |
| **Tianwei Tan** | Beijing University of Chemical Technology, Prof. Member of the Chinese Academy of Engineering |
| **Hailing Tu** | General Research Institute For Nonferrous Metals, Prof. Member of the Chinese Academy of Engineering |
| **Xieqing Wang** | Sinopec Research Institute of Petroleum Processing, Prof. Member of the Chinese Academy of Engineering |
| **Xuguang Wang** | Beijing General Research Institute of Mining and Metallurgy, Dr. Member of the Chinese Academy of Engineering |
| **Dianzuo Wang** | Chinese Academy of Engineering, Dr. Member of the Chinese Academy of Engineering |
| **Haizhou Wang** | China Iron & Steel Research Institute Group, Prof. Member of the Chinese Academy of Engineering |
| **Jingkang Wang** | Tianjin University, Prof. Member of the Chinese Academy of Engineering |
| **Yide Wang** | Taiyuan Iron & Steel (Group) Co., Ltd., Prof. Member of the Chinese Academy of Engineering |
| **Yuqing Weng** | The Chinese Society for Metals, Prof. Member of the Chinese Academy of Engineering |
| **Weizu Wu** | An Institute of Headquarters of General Staff of PLA, Prof. Member of the Chinese Academy of Engineering |

| | |
|---|---|
| **Yicheng Wu** | Technical Institute of Physics and Chemistry, CAS, Prof. Member of the Chinese Academy of Engineering |
| **Chengen Xu** | Sinopec Engineering Incorporation, Prof. Member of the Chinese Academy of Engineering |
| **Delong Xu** | Xi´an University of Architecture and Technology, Dr. Member of the Chinese Academy of Engineering |
| **Huibin Xu** | Beihang University, Dr. Member of the Chinese Academy of Engineering |
| **Kuangdi Xu** | Chinese Academy of Engineering, Prof. Member of the Chinese Academy of Engineering |
| **Qiye Yang** | Sinopec Engineering Incorporation, Prof. Member of the Chinese Academy of Engineering |
| **Guomao Yin** | Steel Tube Research Institute of Ansteel group, Prof. Member of the Chinese Academy of Engineering |
| **Ruiyu Yin** | China Iron & Steel Research Institute Group, Prof. Member of the Chinese Academy of Engineering |
| **Qingtang Yuan** | China Petrochemical Corporation, Dr. Member of the Chinese Academy of Engineering |
| **Weikang Yuan** | East China University of Science and Technology, Dr. Member of the Chinese Academy of Engineering |
| **Guocheng Zhang** | General Research Institute For Nonferrous Metals, Prof. Member of the Chinese Academy of Engineering |
| **Shourong Zhang** | Wuhan Iron & Steel ( Group ) Co., Prof. Member of the Chinese Academy of Engineering |
| **Wenhai Zhang** | China Nerin Engineering Co., Ltd., Prof. Member of the Chinese Academy of Engineering |

| | |
|---|---|
| **Xingdong Zhang** | National Engineering Research Center for Biomaterials, Sichuan University, Prof. Member of the Chinese Academy of Engineering |
| **Liancheng Zhao** | Harbin Institute of Technology, Prof. Member of the Chinese Academy of Engineering |
| **Lian Zhou** | Northwest Institute for Nonferrous Metal Research, Dr. Member of the Chinese Academy of Engineering |
| **Yu Zhou** | Harbin Institute of Technology, Dr. Member of the Chinese Academy of Engineering |
| **Tieyong Zuo** | Beijing University of Technology, Prof. Member of the Chinese Academy of Engineering |
| **Tao Qu** | Kunming University of Science and Technology, Prof. |
| **Ji Yang** | Chinese Academy of Engineering, Prof. |
| **Zuming Liu** | Central South University, Prof. |
| **Guocheng Yao** | Chinese Academy of Engineering, Prof. |
| **Lei Zhao** | China Iron & Steel Research Institute Group, Prof. |
| **Junbo Gong** | Tianjin University, Prof. |
| **Guorui Duan** | Taiyuan Iron & Steel (Group) Co., Ltd., Prof. |
| **Jiawei Yang** | Xi´an University of Architecture and Technology, Prof. |
| **Jidong Guo** | Chinese Academy of Engineering, Prof. |
| **Xuxiao Zhang** | China Iron & Steel Research Institute Group, Prof. |
| **Wenhao Zhang** | Wuhan Iron & Steel ( Group ) Co., Prof. |
| **Xiaoqing Zhan** | China Nerin Engineering Co., Ltd., Prof. |
| **Xuan Zhang** | National Engineering Research Center for Biomaterials, Sichuan University, Prof. |
| **Hongkang Zhu** | Northwest Institute for Nonferrous Metal Research, Prof. |
| **Yufeng Wu** | Beijing University of Technology, Prof. |
| **Xueliang Zhao** | Sinopec Engineering Incorporation, Prof. |

| | |
|---|---|
| **Ying Chen** | Beijing Normal University, Prof. |
| **Peiyong Qin** | Beijing University of Chemical Technology, Prof. |
| **Biqiang Chen** | Beijing University of Chemical Technology, Prof. |
| **Xu Zhang** | Beijing University of Chemical Technology, Prof. |
| **Jing Liu** | Technical Institute of Physics and Chemistry, Chinese Academy of Sciences, Prof. |
| **Zheshuai Lin** | Technical Institute of Physics and Chemistry, Chinese Academy of Sciences, Prof. |
| **Hongwei Gao** | Technical Institute of Physics and Chemistry, Chinese Academy of Sciences, Prof. |
| **Jun Long** | Sinopec Research Institute of Petroleum Processing, Prof. |
| **Jinbiao Guo** | Sinopec Research Institute of Petroleum Processing, Prof. |
| **Jun Fu** | Sinopec Research Institute of Petroleum Processing, Prof. |
| **Songbai Tian** | Sinopec Research Institute of Petroleum Processing, Prof. |
| **Wei Wu** | Sinopec Research Institute of Petroleum Processing, Prof. |
| **Xuhong Mu** | Sinopec Research Institute of Petroleum Processing, Prof. |
| **Zhenyu Dai** | Sinopec Research Institute of Petroleum Processing, Prof. |
| **Shuandi Hou** | Sinopec Research Institute of Petroleum Processing, Prof. |
| **Longpeng Cui** | Sinopec Research Institute of Petroleum Processing, Prof. |
| **Wei Cheng** | Sinopec Research Institute of Petroleum Processing, Prof. |
| **Xiangbo Guo** | Sinopec Research Institute of Petroleum Processing, Prof. |
| **Baoji Zhang** | Sinopec Research Institute of Petroleum Processing, Prof. |
| **Xiahui Gui** | China University of Mining and Technology, Prof. |
| **Xiaokang Yan** | China University of Mining and Technology, Prof. |
| **Guixia Fan** | China University of Mining and Technology, Prof. |
| **Keji Wan** | China University of Mining and Technology, Prof. |
| **Guosheng Li** | China University of Mining and Technology, Prof. |

| | |
|---|---|
| **Ai Wang** | China University of Mining and Technology, Prof. |
| **Yaowen Xing** | China University of Mining and Technology, Prof. |
| **Xiaoyan Kong** | China University of Mining and Technology, Prof. |
| **Zhenyong Miao** | China University of Mining and Technology, Prof. |
| **Zhijun Zhang** | China University of Mining and Technology, Beijing, Prof. |
| **Gen Huang** | China University of Mining and Technology, Beijing, Prof. |
| **Guihong Han** | Zhengzhou University, Prof. |
| **Yanfang Huang** | Zhengzhou University, Prof. |
| **Songtao Huang** | General Research Institute For Nonferrous Metals, Prof. |
| **Shigang Lu** | General Research Institute For Nonferrous Metals, Prof. |
| **Qiang Zhu** | General Research Institute For Nonferrous Metals, Prof. |
| **Ligen Wang** | General Research Institute For Nonferrous Metals, Prof. |
| **Xiumin Chang** | General Research Institute For Nonferrous Metals, Prof. |
| **Jiankang Wen** | General Research Institute For Nonferrous Metals, Prof. |
| **Xujun Mi** | General Research Institute For Nonferrous Metals, Prof. |
| **Lijun Jiang** | General Research Institute For Nonferrous Metals, Prof. |
| **Xiaokui Che** | General Research Institute For Nonferrous Metals, Prof. |
| **Yongsheng Song** | General Research Institute For Nonferrous Metals, Prof. |
| **Zhimin Yang** | General Research Institute For Nonferrous Metals, Prof. |
| **Tengfei Li** | General Research Institute For Nonferrous Metals, Prof. |
| **Qian Zhang** | China Iron & Steel Research Institute Group, Prof. |
| **Huaizhou Cui** | China Iron & Steel Research Institute Group, Prof. |
| **Zhibin Liu** | China Iron & Steel Research Institute Group, Prof. |
| **Jintao Tong** | China Iron & Steel Research Institute Group, Prof. |
| **Chao Tang** | China Iron & Steel Research Institute Group, Prof. |
| **Shuo Feng** | China Iron & Steel Research Institute Group, Prof. |
| **Wenwen Shen** | China Iron & Steel Research Institute Group, Prof. |

| | |
|---|---|
| **Ziyang Zhang** | China Iron & Steel Research Institute Group, Prof. |
| **Qingyuan Tan** | China Iron & Steel Research Institute Group, Prof. |
| **Xiaoqing Xu** | China Iron & Steel Research Institute Group, Prof. |
| **Zhengyan Zhang** | China Iron & Steel Research Institute Group, Prof. |
| **Xiaoqiang Shi** | China Iron & Steel Research Institute Group, Prof. |
| **Yaxin Ma** | China Iron & Steel Research Institute Group, Prof. |
| **Yu Liu** | China Iron & Steel Research Institute Group, Prof. |
| **Linhao Xu** | China Iron & Steel Research Institute Group, Prof. |
| **Feng Kang** | China Iron & Steel Research Institute Group, Prof. |
| **Xinchao Fan** | China Iron & Steel Research Institute Group, Prof. |

# Part Ⅲ

# Keynote Speech and Speaker Introduction

# Atomic Level Control of Quantum Material Growth: From Quantized Anomalous Hall Effect to High Temperature Superconductivity

**Qikun Xue**

Tsinghua University, Beijing, China

**Abstract**: Molecular beam epitaxy (MBE) is a powerful technique for preparation of semiconductors and related heterostructures, among which high mobility two dimensional electron gas and multiple quantum well structure forcascade laser are two of the well-known examples. Combining MBE with scanning tunnelling microscopy (STM) and angle resolved photoemission spectroscopy (ARPES) can even push its power to an unprecedented level. In this talk, I would present our results on the discovery of quantized anomalous hall effect in magnetically doped topological insulators and possible route for solving the high temperature superconductivity by using MBE-STM-ARPES.

Qikun Xue received his BSc in Shandong University in 1984, and PhD degree in condensed matter physics from Institute of Physics, The Chinese Academy of Sciences (CAS) in 1994. From 1994 to 2000, he worked as a Research Associate at IMR, Tohoku University, Japan and a visiting assistant Professor at Department of Physics, North Carolina State University, USA. He became a professor at Institute of Physics, CAS in 1999. He was elected into

The Chinese Academy of Sciences in 2005. Since 2005, he has been a professor in Department of Physics, Tsinghua University. From 2010 to 2013, he was the Chair of Department of Physics and the Dean of School of Sciences. He became the Vice President for Research in May 2013, Tsinghua University. He won the TWAS Prize in Physics in 2010. His research interests include scanning tunneling microcopy/spectroscopy, molecular beam epitaxy, low-dimensional and interface-related superconductivity, topological insulators, and quantum size effects in various low-dimensional structures. He has authored/coauthored ~ 360 papers (5 in *Science*, 9 in *Nature* associated journals, and 31 in *Phys. Rev. Lett.*) with a citation of ~ 7400 times. He has presented more than 100 invited/keynote/plenary talks at international meetings/conferences.

# Regenerative Engineering, a New Field: Theory and Practice

**Cato T. Laurencin**

University of Connecticut, Storrs, Connecticut, USA

**Abstract:** The next ten years we will see unprecedented strides in regenerating musculoskeletal tissues. We are moving from an era of advanced prosthetics, to what I term regenerative engineering. In doing so, we have the capability to begin to address grand challenges in musculoskeletal regeneration. Tissues such as bone, ligament, and cartilage can now be understood from the cellular level to the tissue level. We now have the capability to produce these tissues in clinically relevant forms through tissue engineering techniques. Our improved ability to optimize engineered tissues has occurred in part due to an increased appreciation for advanced materials science/nanotechnology and stem cell technology two relatively new tools for the engineer. Critical parameters impact the design of novel matrices for tissue regeneration. Besides advances in our understanding of the capability of materials, understanding of how cellular and intact tissue behavior can be modulated by material designs will be important. Design of systems for regeneration must take place with a holistic and comprehensive approach, understanding the contributions of cells, biological factors, scaffolds and morphogenesis.

**Cato T. Laurencin** M. D., Ph. D. is an elected member of the U. S. National Academy of Engineering and an elected member of the U. S. Institute of Medicine of the National Academy of Sciences. Professor Laurencin is a University Professor at the University of Connecticut (the 7$^{th}$ in the institution's 130 year history) and serves as the Chief Executive Officer of the Connecticut Institute for Clinical and Translational Science at UCONN. He is the Albert and Wilda Van Dusen Distinguished Endowed Professor of Orthopaedic Surgery, Professor of Chemical, and Biomolecular Engineering, Professor of Materials Engineering, and Professor of Biomedical Engineering at the school. Dr. Laurencin is the Founder and Director of both the Institute for Regenerative Engineering and the Raymond and Beverly Sackler Endowed Center for Biomedical, Biological, Physical and Engineering Sciences at the University of Connecticut. He is a pioneer in the new field of Regenerative Engineering, and his research involves advanced biomaterials science, nanotechnology, stem cell science, and tissue regeneration. He is a Fellow of the Materials Research Society, an International Fellow in Biomaterials Science and Engineering, a Fellow of the Biomedical Engineering Society and was named one of the 100 engineers of the Modern Era by the American Institute of Chemical Engineers at its centennial celebration. Dr. Laurencin was honored by President Bill Clinton with the Presidential Faculty Fellow Award, and honored by President Barack Obama with the Presidential Award of Excellence in Mentoring. Dr. Laurencin earned his B. S. E. degree in Chemical Engineering from Princeton University and his M. D., Magna Cum Laude from the Harvard Medical School. He earned his Ph. D. in Biochemical Engineering/Biotechnology from the Massachusetts Institute of Technology where he was named a Hugh Hampton Young Fellow.

# Simulation and Modelling of Particulate Systems

**Aibing Yu**

Monash University, Melboume, Victoria, Australia

**Abstract:** Particle science and technology is a rapidly developing interdisciplinary research area with its core being the understanding of the relationships between micro- and macro-scopic properties of particulate/granular matter-a state of matter that is widely encountered but poorly understood. The macroscopic behaviour of particulate matter is controlled by the interactions between individual particles as well as interactions with surrounding gas or liquid and wall. Understanding the microscopic mechanisms in terms of these interaction forces is therefore key to leading to truly interdisciplinary research into particulate matter and producing results that can be generally used. This aim can be effectively achieved via particle scale research based on detailed microdynamic information such as the forces acting on and trajectories of individual particles in a considered system. In recent years, such research has been rapidly developed worldwide, mainly as a result of the rapid development of discrete particle simulation technique and computer technology. This talk will present an overview of the work in this direction. It covers the theoretical developments and case studies under different conditions. It is demonstrated through representative examples in chemical, metallurgical and materials engineering that the study of small particles is well linked to many challenging problems in big science. The examples also demonstrate that particle scale approach has gradually emerged to be a powerful tool not only for fundamental research but also for engineering application. Finally, areas for future development are briefly discussed.

**Aibing Yu** specialized in process metallurgy, obtaining BEng in 1982 and MEng in 1985 from Northeastern University, PhD in 1990 from the University of Wollongong, and DSc in 2007 from the University of New South Wales (UNSW). He has been with UNSW School of Materials Science and Engineering since 1992. Currently he is Scientia Professor, leading a world-class research facility "Simulation and Modelling of Particulate Systems" (SIMPAS). He is also Director of Australia-China Joint Research Centre for Minerals, Metallurgy and Materials. As from April 2014, he will join Monash University as its Pro Vice Chancellor and President of Monash-Southeast University Joint Research Institute. He is a world-leading scientist in particle/powder technology and process engineering, has authored >750 publications including >430 articles collected in the ISI Web of Science, and delivered many plenary/keynote presentations at different international conferences. He is Editor-in-Chief (2014 –), Handbook (a book series) of *Powder Science and Engineering*, and Editor, *Powder Technology* (2013–) and *Particuology* (2008–2013), and has been on the editorial or advisory board of about 20 learned journals including *Industrial & Engineering Chemistry Research*, *Powder Technology*, *Granular Matter*, *ISIJ International*, and *National Science Review*. He is a recipient of various prestigious fellowships and awards, including ARC Queen Elizabeth II, Australian Professorial and Federation Fellowships, the Josef Kapitan Award from the Iron and Steel Society, Ian Wark Medal from Australian Academy of Science, ExxonMobile Award from Australian and New Zealand Federation of Chemical Engineers, NSW Scientist of Year in engineering, mathematics and computer science, and Top 100 Most Influential Engineers in Australia. He is an elected Fellow of both Australian Academy of Science (AAS) and Australian Academy of Technological Sciences and Engineering (ATSE).

# Chemical Looping Technology: Iron-Based Ohio State Processes

**Dikai Xu, Liang-Shih Fan** *

William G. Lowrie Department of Chemical and Biomolecular Engineering, The Ohio State University, Columbus, Ohio, USA

**Abstract:** In recent years, chemical looping processes have been widely studied as a novel method for carbon capture during the conversion of carbonaceous fuels. Chemical looping processes indirectly combust carbonaceous fuels using oxygen carriers, and $CO_2$ is captured *in-situ* without using traditional post-combustion capture technologies. The Ohio State University (OSU) is developing a unique iron-based chemical looping technology that enables maximum conversion of oxygen carrier and fuels. Thermodynamic analysis shows that with full fuel conversion the counter-current moving bed reactor can achieve an oxygen carrier conversion that is far greater than other types of reactors, while also producing high purity $H_2$ through the steam-iron reaction. To date, the syngas chemical looping (SCL) and the coal direct chemical looping (CDCL) have been successfully demonstrated on two 25 $kW_{th}$ sub-pilot units.

## 1 Introduction

With the growing concern of global climate change due to greenhouse gas emissions, chemical looping processes have been widely studied due to its capability of efficiently converting fuels with zero $CO_2$ emission[1]. In chemical looping processes, a solid oxygen carrier, usually metal oxides, is employed to convert carbonaceous fuels into a stream consisting of carbon dioxide ($CO_2$) and water ($H_2O$) or hydrogen ($H_2$)

* Corresponding author, E-mail: fan.1@osu.edu.

and carbon monoxide (CO) in the fuel reactor, thus avoiding the mixing of the fuel and air eliminating the need for further $CO_2$ separation[2]. The reduced oxygen carrier is regenerated in either a single reactor using air, which generates heat that can be used for electricity generation, or two reactors using steam for partial re-oxidation and hydrogen generation and using air for full oxidation.

Recently, worldwide research has explored various aspects of chemical looping, and several different types of oxygen carriers and reactors have been proposed[2]. Most of the oxygen carriers consist of an oxide of Ni, Fe, or Cu as the active compound and supported by inert materials such as $Al_2O_3$. Although nickel-based oxygen carriers are reactive, they suffer from a high cost and the toxicity of NiO. Moreover, thermodynamics of nickel-based oxygen carriers results in a high level of CO in the effluent stream that requires further treatment. Copper-based oxygen carriers decompose and release $O_2$ at high temperature, and hence exhibit good reactivity. However, due to the low melting point of copper, copper-based oxygen carriers suffer from agglomeration and defluidization in fluidized bed reactors. This is usually overcome by adopting a low CuO mass ratio in the oxygen carrier formula and sacrificing the oxygen carrying capacity of the material. Iron-based oxygen carriers are more economical and environmentally benign. The thermodynamic properties of iron allow for a much lower CO level in the effluent stream as compared to nickel-based oxygen carriers. A proper selection of support material is required to overcome their deficit in reactivity. Circulating Fluidized Beds (CFB) are widely employed as the Fuel reactor because of their ability to handle high solids circulation and excellent gas-solid contact. However, in a fluidized bed reactor using iron-based oxygen carriers, $Fe_2O_3$ can only be reduced to $Fe_3O_4$ due to thermodynamic limitations, which corresponds to 11% of the oxygen carrying capacity being utilized.

The Ohio State University (OSU) is developing a unique iron-based chemical looping technology capable of fully converting solid or gaseous fuels into $CO_2$ and steam. As shown in Fig.1, the OSU chemical looping technology consists of a reducer, an oxidizer, and a combustor. The major reaction in each of the reactors are:

Reducer: $$C_xH_yO_z+Fe_2O_3 \longrightarrow CO_2+H_2O+Fe/FeO \quad (1)$$

Oxidizer: $$Fe/FeO+H_2O \longrightarrow Fe_3O_4+H_2 \quad (2)$$

Combustor: $$Fe_3O_4+O_2 \longrightarrow Fe_2O_3+Q \quad (3)$$

Overall reaction: $$C_xH_yO_z+H_2O+O_2 \longrightarrow CO_2+H_2O+H_2+Q \quad (4)$$

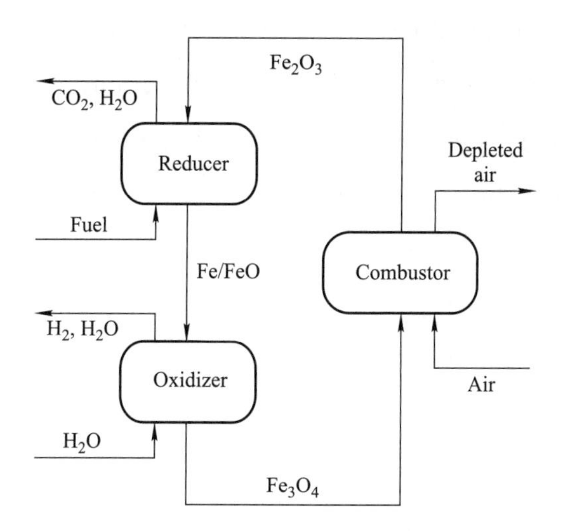


**Fig.1 Diagram of OSU chemical looping system**

The gaseous product stream exiting the reducer is predominantly $CO_2$ after condensing out the steam, which is ready for sequestration or utilization. Meanwhile, high purity $H_2$ is produced from the oxidizer, and heat is recovered from the combustor. The reducer and oxidizer are both counter-current moving bed reactors. The unique feature of moving bed reactors maximizes the conversion of fuel, steam, and oxygen carrier[2-4].

To date, OSU has designed and constructed a 25 $kW_{th}$ syngas chemical looping (SCL) sub-pilot unit and a 25 $kW_{th}$ coal direct chemical looping (CDCL) sub-pilot unit for full oxidation of gaseous and solid fuel, respectively, on which a total of more than 850 hours of integrated operation with high fuel conversion has been demonstrated[5-8]. The SCL unit has achieved the co-generation of >99. 99% $H_2$ while generating a pure $CO_2$ stream in syngas combustion. The CDCL unit has demonstrated the conversion of several types of solid fuels including biomass, coal, and metallurgical coke.

## 2 Oxygen carrier development

In chemical looping processes, the oxygen carrier circulates between the reactors, conveying oxygen between air/steam and fuels via cyclic oxidation-reduction reactions. Therefore, a desirable oxygen carrier will maintain its mechanical integrity and chemical reactivity over several cycles, minimizing costs by reducing the oxygen carrier make-up rate. Widely studied oxygen carrier materials include $CaSO_4$, the oxides of Ni, Cu, Fe, and Mn. Solid supports include $Al_2O_3$ and $MgAl_2O_4$ in order to improve the reactivity and mechanical strength[9].

OSU has tested more than 600 oxygen carrier materials based on different metal oxides, support material, and synthesis methods, and a series of $Fe_2O_3$ based composite oxygen carriers with extraordinary reactivity and recyclability has been developed. As shown in Fig. 2, the oxygen carrier maintains reactivity for more than 100 reduction-oxidation cycles, while pure $Fe_2O_3$ loses its capacity within a few cycles. Compared to other oxygen carrier materials, iron-based oxygen carriers have desirable thermodynamic properties: the oxidized state ($Fe_2O_3$) is capable of fully converting fuels to $CO_2$ and $H_2O$, while the reduced states (Fe/FeO) can react with steam to generate $H_2$.

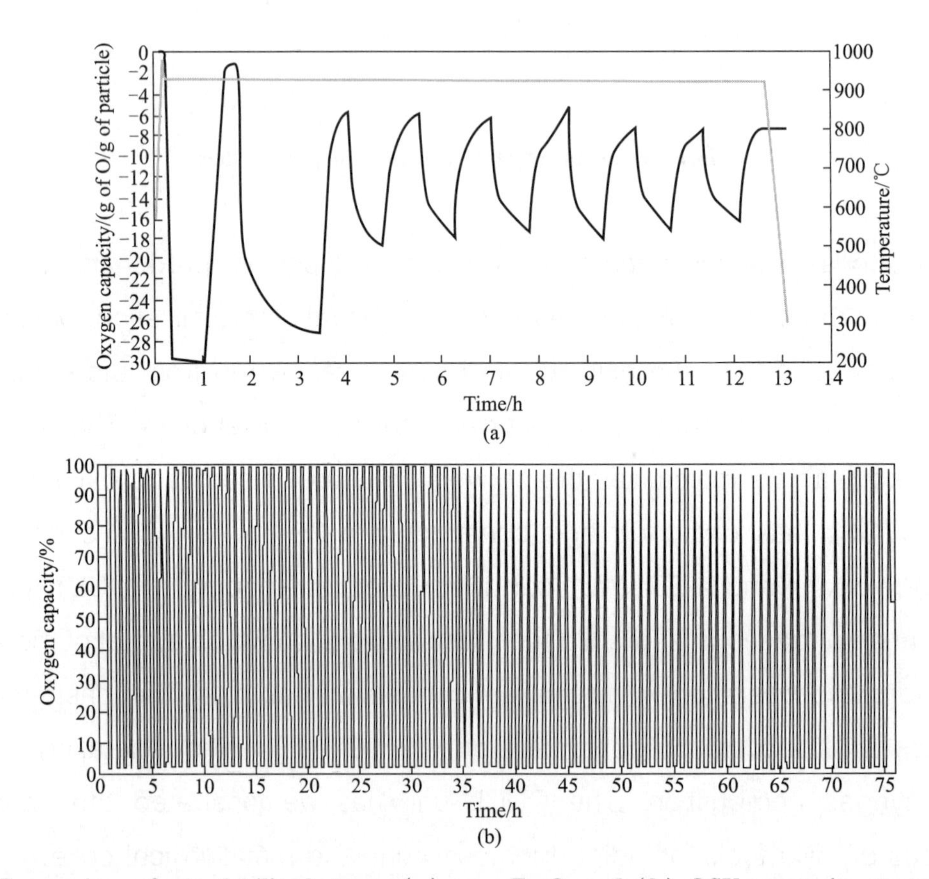


**Fig.2 Comparison of recyclability between (a) pure $Fe_2O_3$ and (b) OSU composite oxygen carrier**

## 3 Iron-based chemical looping reactor design and operation

To date, multiple demonstration projects for chemical looping technologies have been constructed, and the total operation time has well exceeded 2000 hours[10-13]. In the vast majority of these projects, fluidized bed reducers (fuel reactors) are used. The OSU chemical looping technology employs the unique counter-current moving bed reducer and oxidizer, which maximizes the oxygen carrier conversion while fully oxidizing the fuels[14].

If only the thermodynamics in the reactors are considered, the behavior of a gas-solid fluidized bed reactor coincide with that of a co-current moving bed reactor, namely, the gaseous products are in chemical equilibrium with the solid products. Thus, the gas-solid contact mode of reducer and oxidizer can be classified into two modes: mode 1 for fluidized bed or co-current moving bed reactors, and mode 2 for counter-current moving bed reactors[14,15]. For each mode, specific thermodynamic limitations exist between the oxygen carrier and fuel conversion.

Fig.3 considers the chemical equilibrium of $FeO_x$ reacting with $H_2/H_2O$ at 850℃. The solid lines represent the equilibrium states: vertical lines correspond to the equilibrium between two solid phases and the gas phase with a certain composition. The horizontal lines correspond to the equilibrium between one solid phase and the gas phase with a flexible composition. The operation of the reducer falls in the lower left region of the diagram, while that of the oxidizer falls in the upper right region.

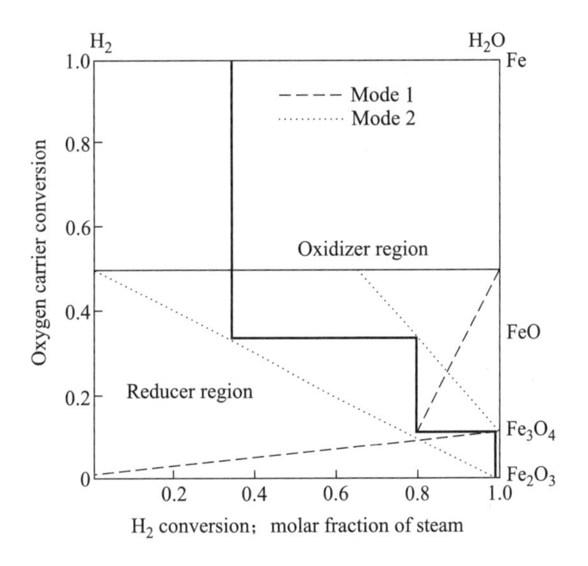


**Fig.3 Operating line for reducer and oxidize**

An oxygen material balance shows that the operating lines of the reducer and the oxidizer are line segments in the diagram [14]. Along the direction of gas/solid flow, the conversions of gas and oxygen carrier vary along the operating line. The slope of the operating line is proportional to the ratio between gas and oxygen carrier flow. The operating lines for mode 1 have positive slopes and that for mode 2 have negative slopes[14].

The operating lines for both modes are shown in Fig. 3 with dashed lines representing mode 1 and dotted lines representing mode 2. As shown in the reducer region of Fig.3, the oxygen carrier conversion for full conversion of $H_2$ when operating in mode 1 is only 11. 1%, corresponding to the reduction of $Fe_2O_3$ to $Fe_3O_4$. However, in mode 2, an oxygen carrier conversion of about 50% can be achieved, and $Fe_2O_3$ is reduced to a mixture of Fe and FeO. Therefore, to convert a fixed amount of fuel, the solid circulation rate of mode 2 is only about 20% of mode 1. Moreover, in mode 1, the reduced oxygen carrier ($Fe_3O_4$) cannot react with steam in the oxidizer to produce $H_2$.

When mode 2 is adopted in the reducer, the reduced oxygen carrier can be oxidized by steam to generate $H_2$ in the oxidizer. The operating line for the oxidizer is shown in the oxidizer region in Fig.3. For the same oxygen carrier conversion (from Fe/FeO to $Fe_3O_4$), mode 1 produces a mixture of $H_2$ and $H_2O$ with about 20% $H_2$ concentration, while $H_2$ concentration in mode 2 is about 35%. Thus, the steam flow rate and energy cost required for mode 2 is much lower. A steam conversion greater than 50% can be achieved if the operating temperature of the oxidizer can be further lowered.

For iron-based chemical looping processes, using a counter-current moving bed reducer and oxidizer is advantageous over fluidized beds and co-current moving bed reactors. OSU has designed and constructed demonstration units to further study the counter-current moving bed reactors in chemical looping processes.

## 4 Iron-based chemical looping processes

### 4.1 Syngas chemical looping (SCL) process

SCL technology converts gaseous fuels including syngas or natural gas into electricity and $H_2$ with zero $CO_2$ emission. A schematic of the SCL proces is shown in Fig.4[2]. Syngas, which can be generated by coal gasification, is sent to the bottom of the counter-current moving bed reducer after gas clean-up. The energy in the high temperature gas stream from the reactors can be recovered for electricity generation. $H_2$ produced in the process can be used in fuel cell systems or downstream chemical production.

OSU has constructed and operated a 25 $kW_{th}$ SCL sub-pilot unit, which has achieved high purity $H_2$ production and complete carbon capture in syngas combustion[5,6]. A continuous 3-day operation with steady generation of high purity $CO_2$ and $H_2$ was achieved. Operation data showed nearly complete conversion of syngas in the reducer and >99. 99% $H_2$ purity in the product stream from the oxidizer[6]. The

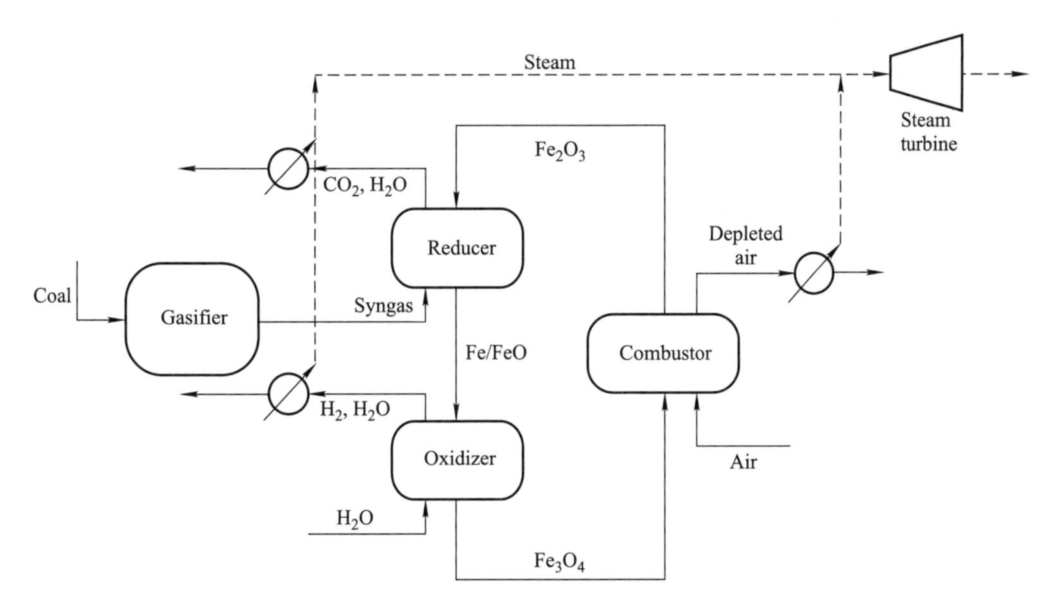


**Fig.4 Diagram of syngas chemical looping system**

combustion of methane has also been performed, achieving a methane conversion greater than 99. 5%. The success in operating the sub-pilot unit confirmed the feasibility of the OSU chemical looping technology.

Currently, OSU is conducting a 250 $kW_{th}$ pilot scale pressurized SCL process study at the National Carbon Capture Center (NCCC). The unit is designed to operate at 10 atm combusting the syngas from a Kellogg Brown & Root (KBR) gasifier. The construction of the SCL pilot unit is complete and operations will start by the end of 2014.

## 4. 2 Coal direct chemical looping (CDCL) process

The OSU CDCL technology can directly convert solid fuels such as coal, biomass, and metallurgical coke in the reducer, eliminating the need for a coal gasifier. The CDCL process consists of a two-stage counter-current moving bed reactor and a fluidized bed combustor. Pulverized solid fuels are introduced and thermally devolatilized at the middle of the reducer, by which the reactor is divided into two stages. The gaseous volatiles flow upward to the upper stage of the reactor, and is fully oxidized to $CO_2$ and $H_2O$ by $Fe_2O_3$(Fig.5). The solid residuals from devolatilization move downwards with the oxygen carrier into the lower stage, where a small stream of $CO_2$ and/or $H_2O$ is introduced to promote the gasification of the residuals[7,16].

OSU has developed a 2. 5 $kW_{th}$ bench scale moving bed reducer and a 25 $kW_{th}$ sub-pilot integrated unit to study the conversion of metallurgical coke, lignite, sub-

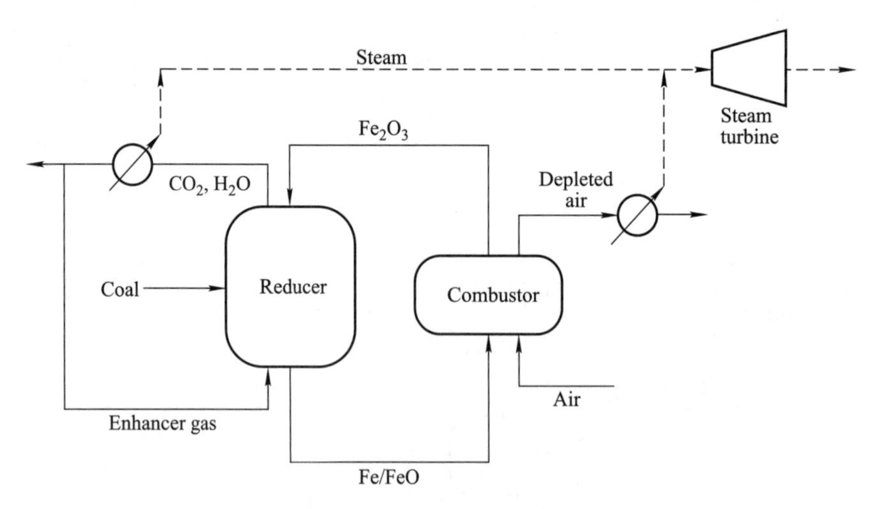


**Fig.5 Coal direct chemical looping ( CDCL ) process flow diagram**

bituminous coal, bituminous coal, anthracite, and woody biomass[7,8,17-19]. Conversion studies, totaling more than 300 hours, of various volatiles and solid fuels have been conducted on the 2. 5 $kW_{th}$ bench scale unit.

The 25 $kW_{th}$ sub-pilot unit fully employs non-mechanical structures for gas solid flow control. More than 550 hours of operation have been performed. The longest duration for continuous operation lasted over 200 hours, which is the longest continuous demonstration of chemical looping technology for solid fuel conversion[9]. Steady operation with well-controlled solid circulation and gas flow was maintained throughout the operation, and the capability of converting different flow rates of PRB coal and lignite was demonstrated. The conversion of both kinds of coals in the sub-pilot unit were greater than 96%, with a $CO_2$ stream of 99. 4% to 99. 8% purity generated from the reducer. The concentration of byproducts including $CH_4$ and CO were lower than 0. 5%. The results confirm that the reducer is capable of fully oxidizing the solid fuels. Minimal $CO_2$ is detected at the outlet of the combustor, indicating a minimal carbon carry-over from the reducer to the combustor.

### 4. 3 Integration of chemical looping technology and other technologies

OSU's chemical looping processes are highly flexible to accept different types of fuels and output desired products such as electricity, $H_2$, and liquid fuels. Fig.6 illustrates the integration of the OSU SCL technology and the conventional coal to liquid ( CTL) process[2,15]. The SCL process can be integrated into an existing CTL system to improve process efficiency while lowering $CO_2$ emissions.

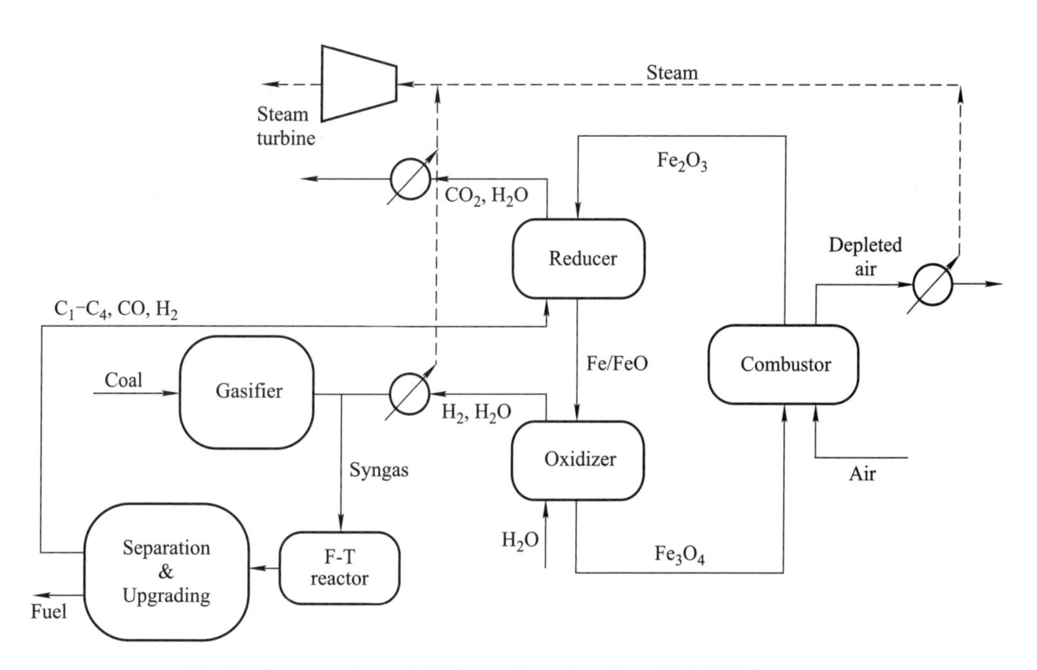


**Fig.6 Integration of SCL and coal to liquid (CTL) process**

In the integration, $H_2$ from the SCL oxidizer is used to adjust the syngas composition to the appropriate $H_2$ : CO ratio necessary for the Fischer-Tropsch process. Light hydrocarbons from the Fischer-Tropsch process and product refinery are used as the fuel of the SCL process, and be converted to a sequestration-ready $CO_2$ stream. The steam produced in the Fischer-Tropsch reactor can be conveniently used as the feedstock of the SCL oxidizer. Therefore, the SCL process substitutes the water-gas shift and $CO_2$ removal units in the conventional CTL process. An independent analysis by US DOE shows that compared to the conventional CTL process, the integration can increase liquid yield by 10% and reduce $CO_2$ emissions by 19%[20].

# 5 Techno-economic analysis of chemical looping technology

A series of techno-economic analyses on the OSU chemical looping technology shows that compared to current processes with $CO_2$ capture, chemical looping improves the efficiency and lowers the cost. The efficiency of SCL technology, shown in Fig.4, is 5.5% higher than a conventional coal gasification process for $H_2$ production with 90% $CO_2$ capture. If the $H_2$ is converted to electricity with 60% efficiency (HHV), the levelized cost of electricity (LCOE) is 10% lower[21].

The CDCL process for electricity generation is advantageous over a conventional pulverized coal (PC) power plant with carbon capture. For a 550 $MW_e$ power plant, the efficiency of the CDCL process with 100% $CO_2$ capture is 23. 5% higher than a PC power plant capturing 90% $CO_2$ by MEA, and the cost electricity is only 33% higher than a PC power plant without $CO_2$ capture, compared to a 71% cost increase for 90% $CO_2$ capture with MEA in a PC power plant[22]. Therefore, the CDCL process is capable of achieving the US DOE target of less than 35% cost of electricity increase with 90% $CO_2$ capture.

## 6 Conclusion

OSU has developed unique iron-based chemical looping processes using counter-current moving bed reactors, enabling the efficient conversion of carbonaceous fuels with zero $CO_2$ emissions. Over 600 oxygen carrier materials have been synthesized and tested. The current iron-based composite oxygen carrier can maintain excellent reactivity for more than 100 cycles. The performance of the oxygen carrier is further confirmed in the bench and sub-pilot scale demonstration units. Based on the thermodynamic property of the iron-based oxygen carrier, OSU designed the unique counter-current moving bed reducer and oxidizer, which maximizes the oxygen carrier conversion while ensuring full fuel conversion. Compared to a fluidized bed reducer, the solid circulation rate for the counter-current moving bed reducer is about 80% smaller, and $Fe_2O_3$ can be reduced to a lower oxidation state, enabling $H_2$ generation in the oxidizer. Therefore, the chemical looping process can be integrated with a variety of technologies such as CTL to achieve a higher energy efficiency.

The OSU chemical looping technology has been successfully demonstrated on two 25 $kW_{th}$ sub-pilot units. The SCL sub-pilot unit achieved >99. 99% purity $H_2$ production with complete $CO_2$ capture in syngas combustion. A 250 $kW_{th}$ pressurized SCL pilot unit has been constructed at NCCC and will start operation by the end of 2014. The CDCL sub-pilot unit employs a two-stage counter-current moving bed reducer and non-mechanical structures for controlling gas solid flow. Long term continuous operation with stable gas and solid flow has been demonstrated on the unit. The conversion of various solid fuels including coke, coal, and biomass has been confirmed for over 550 hours of operation.

## Acknowledgement

The helpful technical and editorial assistance from William Wang is gratefully acknowledged. The work was supported by the C. John Easton Endowed Funds from The Ohio State University.

## References

[1] Figueroa J D, Fout T, Plasynski S, et al. Advances in $CO_2$ capture technology—The US Department of Energy's Carbon Sequestration Program [J]. International Journal of Greenhouse Gas Control, 2008, 2(1): 9-20.

[2] Fan L S. Chemical looping systems for fossil energy conversions [M]. New York: Wiley, 2010.

[3] Li F, Fan L S. Clean coal conversion processes-progress and challenges [J]. Energy & Environmental Science, 2008, 1(2): 248-267.

[4] Thomas T J, Fan L S, Gupta P, et al. Combustion looping using composite oxygen carriers: U. S. Patent 7,767, 191[P]. 2010-8-3.

[5] Sridhar D, Tong A, Kim H, et al. Syngas chemical looping process: Design and construction of a 25 $kW_{th}$ subpilot unit [J]. Energy & Fuels, 2012, 26(4): 2292-2302.

[6] Tong A, Sridhar D, Sun Z, et al. Continuous high purity hydrogen generation from a syngas chemical looping 25 $kW_{th}$ sub-pilot unit with 100% carbon capture [J]. Fuel, 2012, 103: 495-505.

[7] Kim H R, Wang D, Zeng L, et al. Coal direct chemical looping combustion process: Design and operation of a 25-$kW_{th}$ sub-pilot unit [J]. Fuel, 2013, 108: 370-384.

[8] Bayham S C, Kim H R, Wang D, et al. Iron-based coal direct chemical looping combustion process: 200-h continuous operation of a 25-$kW_{th}$ subpilot unit[J]. Energy & Fuels, 2013, 27(3): 1347-1356.

[9] Fan L S, Li F. Chemical looping technology and its fossil energy conversion applications [J]. Industrial & Engineering Chemistry Research, 2010, 49(21): 10200-10211.

[10] Adanez J, Abad A, Garcia-Labiano F, et al. Progress in chemical-looping combustion and reforming technologies[J]. Progress in Energy and Combustion Science, 2012, 38(2): 215-282.

[11] Andrus H E, Burns G, Chiu J H, et al. Hybrid combustion-gasification chemical looping coal power technology development Phase III final report, Alstom Power Inc[R]. PPL-08-CT-25, Contract DE-FC26-03NT41866, US Department of Energy, National Energy Technology Laboratory, 2006.

[12] Kolbitsch P, Bolhar-Nordenkampf J, Pröll T, et al. Comparison of two Ni-based oxygen carriers for chemical looping combustion of natural gas in 140 kW continuous looping operation [J]. Industrial & Engineering Chemistry Research, 2009, 48(11): 5542-5547.

[13] Shulman A, Linderholm C, Mattisson T, et al. High reactivity and mechanical durability of $NiO/NiAl_2O_4$ and $NiO/NiAl_2O_4/MgAl_2O_4$ oxygen carrier particles used for more than 1000 h in a 10 kW CLC reactor[J]. Industrial & Engineering Chemistry Research, 2009, 48(15): 7400-7405.

[14] Li F, Zeng L, Velazquez-Vargas L G, et al. Syngas chemical looping gasification process: Bench-scale studies and reactor simulations[J]. AIChE Journal, 2010, 56(8): 2186-2199.

[15] Fan L S, Zeng L, Wang W, et al. Chemical looping processes for $CO_2$ capture and carbonaceous fuel conversion-prospect and opportunity [J]. Energy & Environmental Science, 2012, 5(6): 7254-7280.

[16] Li F, Zeng L, Fan L S. Biomass direct chemical looping process: Process simulation [J]. Fuel, 2010, 89(12): 3773-3784.

[17] Luo S, Bayham S, Zeng L, et al. Conversion of metallurgical coke and coal using a coal direct chemical looping (CDCL) moving bed reactor[J]. Applied Energy, 2014, 118(1): 300-308.

[18] Luo S, Majumder A, Chung E, et al. Conversion of woody biomass materials by chemical looping process: Kinetics, light tar cracking, and moving bed reactor behavior [J]. Industrial & Engineering Chemistry Research, 2013, 52(39): 14116-14124.

[19] Zeng L, He F, Li F, et al. Coal-Direct Chemical looping gasification for hydrogen production: Reactor modeling and process simulation [J]. Energy & Fuels, 2012, 26(6): 3680-3690.

[20] Gray D, Klara J, Tomlinson G, et al. Chemical-looping process in a coal-to-liquids configuration: Independent assessment of the potential of chemical-looping in the context of a Fischer-Tropsch plant[R]//Report No.: DOE/NETL-2008/1307. Contract No.: NBCH-C-020039. Sponsored by the US Department of Energy. Pittsburgh P A: National Energy Technology Laboratory, 2007.

[21] Li F, Zeng L, Fan L S. Techno-economic analysis of coal-based hydrogen and electricity cogeneration processes with $CO_2$ capture [J]. Industrial & Engineering Chemistry Research, 2010, 49(21): 11018-11028.

[22] Connell D, Dunkerley M. Techno-economic analysis of a coal direct chemical looping power plant with carbon dioxide capture[C]//Proceedings of the 37th International Technical Conference on Clean Coal and Fuel Systems. Clearwater, USA, 2012: 29-40.

**Dr. L. S. Fan** is Distinguished University Professor and C. John Easton Professor in Engineering in the Department of Chemical and Biomolecular Engineering at The Ohio State University. His research fields are in fluidization, powder technology, and multiphase reaction engineering. Professor Fan is a member of the U. S. National Academy of Engineering, an Academician of Academia Sinica and a Foreign Member of Chinese Academy of Engineering, Australia Academy of Technology Science and Engineering, and Mexican Academy of Sciences. He was named in 2008 as one of the " One Hundred Engineers of the Modern Era" by the AIChE.

# Photocatalysis-based Novel Technologies for Clean Energy and Environment

**Xianzhi Fu**

Fuzhou University

State Key Laboratory of Photocatalysis on Energy and Environment

National Environmental Photocatalysis Engineering Research Center, Fuzhou, Fujian, China

**Abstract**: Energy shortage and environmental pollution are key issues of modern society. Photocatalytic technology can utilize sunlight to drive many relevant chemical reactions, like water splitting for hydrogen production, carbon dioxide reduction and degradation of organic pollutants, and it is therefore one of the ideal approaches for clean energy production and environmental remediation in the future, showing great promise in energy and environment fields. Thus far, the photocatalytic technologies in practical applications are limited by low quantum yield of photocatalytic processes and low utilization efficiency of sunlight. To solve these problems, scientists are carrying out many basic and applied studies, mainly focusing on the design and synthesis of new photocatalysts, modulation of crystal phase/facet, fabrication of heterojunctions and cocatalysts, band structure engineering of photocatalysts, photocataltyic reaction mechanism and so on. The presentation will cover main research progress of these aspects, together with some applied examples of environmental photocatalytic technologies. Some future perspectives of photocatalytic technologies are also discussed.

Today I am going to give a talk on *Photocatalysis-based Novel Technologies for Clean Energy & Environment*. The presentation includes four parts: ① the research background and introduction to photocatalysis, ② the recent research progress on photocatalysis, ③ the applications of photocatalytic technologies, and ④ the prospects.

Firstly,I would like to briefly introduce the research background of photocatalysis. The energy & environment issue has been considered as one of the most challenging concerns that the humanity has to address in 21$^{st}$ century. This has been already mentioned by many times in this morning's plenary lectures. Currently, the world's growing energy demands conflict seriously with the rapid depletion of fossil fuels such as coal,oil and natural gas,and thus the energy issue has become hot research topics. Meanwhile, the environmental problems are becoming more urgent. For example, the soil, atmosphere and water have been seriously polluted. In addition, the extensive consumption of fossil fuels significantly increases the emission of $CO_2$ greenhouse gas, resulting in global warming. This will raise the sea level to induce glaciers disappearance and flood-prone.

The search for means to supply a sufficient amount of renewable energy is considered as an ideal solution to the energy & environment issues. For a long time,the solar energy, as the inexhaustible source of clean energy, has drawn great attention. According to the statistics, the total solar energy reaching on the surface of the Earth approximately equals to the heat released by 130 trillion tons of coal per year, and the sun can provide sufficient energy in one hour that can supply our energy requirements for a year. In recent years,there has been a burst of interest and activity in solar energy utilization, including the solar-to-thermal, solar-to-electric and solar-to-chemical conversions. As a solar-to-chemical conversion pathway, photocatalysis is a recently-developed technology, which has already been used for environmental remediation, energy conversion and $CO_2$ reduction. Water can be photocatalytically splitted into hydrogen and oxygen gases to achieve the solar-to-hydrogen conversion, while organic pollutants in air and water can be decomposed via photocatalytic reactions.

Currently, the application of photocatalysis in environment field has already come into practical use, but the study of photocatalysis in energy and photocatalytic $CO_2$ reduction are still in the laboratory stage. Heterogeneous photocatalysis is essentially a process of photo-induced redox reactions, involving the generation, migration, and separation of photo-generated charge carriers over semiconductors, in which the adsorbed substrates undergo redox reactions with the separated electrons and holes. As a new technology for environmental remediation, photocatalysis can directly utilize the solar energy to purify the environment, such as complete degradation of pollutants at room temperature and effectively killing bacteria and viruses with good safety, no secondary pollution, wide-adaptability and long-term function. As a new technology for energy generation, photocatalysis can directly convert the low-density solar energy into high-density chemical energy without the emission of $CO_2$ and secondary pollution.

To achieve the large-scale application of photocatalysis based on $TiO_2$, a series of key issues related to science and technology is still needed to be solved. First is the serious recombination of photo-induced charge carriers, strongly depending on the structure and property of $TiO_2$, leading to low quantum efficiency. Second is the wide band-gap of $TiO_2$, which means that only the UV light with photon energy larger than 3. 2 eV can be adsorbed by $TiO_2$, leaving the visible-light-dominated portion of the solar spectrum. Thus, $TiO_2$ photocatalyst exhibits a rather low efficiency of solar energy utilization. Additionally, the photocatalysis always involves the gas-solid-liquid-light heterogeneous reactions, which makes the reaction system very complicated. To better address the above problems, a series of fundamental issues needs to be well solved, such as the unraveling of structure-activity relationship, the study on the mechanism of photocatalytic process, and the development of pathways to improve the efficiency of photocatalysis. These now become frontier and hot topics in the field of photocatalysis from the viewpoints covering fundamental research as well as practical applications.

Recently, many systematical and in-depth investigations have been conducted to address the key scientific and technological issues mentioned above, such as the development of highly efficient visible-light-active photocatalysts, mechanism study on heterogeneous photocatalytic reactions, the development of methods to enhance photocatalytic efficiency and the broadening of applications of photocatalysis. Photocatalysis has become one of the hot topics and the scientific publications on photocatalysis grow rapidly, some of which are often reported in the top journals, including *Nature* and *Science*. In the case of research in China, a leapfrog development has been achieved since 2000. According to the statistics of foreign scholars, the number of papers published in international journals on the topic of photocatalysis by Chinese scientists is the largest. Obviously, photocatalysis has become one of the most internationally competitive fields in our country.

Next, I will briefly present the recent progress on photocatalysis. Currently, the central challenge of photocatalysis still concentrates on the development of new photocatalysts, because many of the troublesome issues on photocatalysis could be well addressed if new type of photocatalysts endow with promising properties. Until now, many synthetic strategies such as surface modification, semiconductor hybridizing, molecular design, ionic doping, solid solution formation, quantum size effect, facet or crystal engineering and cocatalyst/heterojunction have been already adopted for the design and preparation of visible-light photocatalysts with high efficiency and high stability. There are two classes of photocatalysts, one is $TiO_2$ -based photocatalysts derived from the modification of pristine $TiO_2$, and the other is non-$TiO_2$ photocatalysts. Here are some examples of the recently developed new photocatalysts. In the case of $TiO_2$-based photocatalysts, Asahi et al. demonstrated that doping of $TiO_2$ with nonmetal N atoms exhibited a good visible-light photocatalytic performance for the degradation of organic compounds and a super hydrophilic performance without abating the UV-light performance. Mao and co-workers used a hydrogenation processing method to induce a large number of crystal lattice defects on $TiO_2$, significantly reducing the band gap energy which turns the white $TiO_2$ to black $TiO_2$. The third example is the facet-controlled synthesis of $TiO_2$. Gaoqing (Max) Lu and Huiming Cheng *et al.* demonstrated that $TiO_2$ single crystal with the predominant crystal surface of {001}, {101} or {010} can be synthesized by using HF under hydrothermal conditions. The relationship of photocatalytic activity with the crystal facets and electronic structures was also carefully studied, providing a useful pathway to enhance photocatalytic performance. For non-$TiO_2$ photocatalysts, the first example is the development of solid solutions as excellent visible-light-active photocatalysts. Domen et al. prepared nitride or oxide based solid solution photocatalysts for solar water splitting by high temperature nitridation methods, such as GaN-ZnO and GeN-ZnO. A quantum efficiency of as high as 2.5% between 420–440 nm was achieved on GaN-ZnO solid solution for photocatalytic overall water splitting. The second example is a metal-free polymeric photocatalyst. Xinchen Wang et al. recently reported that graphite carbon nitride ($g-C_3N_4$), an organic semiconductor, can function as a metal-free polymeric photocatalyst. On the basis of this work, a series of efficient $g-C_3N_4$-based photocatalysts has been synthesized by copolymerization, templating method, Pt metal modification, and heterojunction construction, broadening the research area of photocatalysis from inorganic semiconductors to organic ones. The third one is the metal organic frameworks (MOFs)-based photocatalysts. Zhaohui Li et al. demonstrated that the replace of terephthalic acid link molecules in MIL-125 (Ti) with amino terephthalic acid can induce a visible light absorption in MOF $NH_2$-MIL-125(Ti), which was firstly used as visible light photocatalysts for $CO_2$ reduction.

The study on photocatalytic reaction mechanism also got some important results, such as reaction kinetics and rate-determined steps. For example, the reaction mechanism of photocatalytic selective oxidation of organics on g-$C_3N_4$ had been clearly revealed by Xinchen Wang et al. It revealed that the activation of molecular oxygen by photoinduced electrons to generate superoxide radical anion is the reaction pathway for the selective oxidation of organics. Jiesheng Chen et al. confirmed that the $Zn^+$ in ZSM molecular sieve is the reactive sites for the photocatalytic activation of methane to produce ethane and hydrogen. Our research demonstrated that the Brønsted Acid centers on solid superacids can greatly improve the photocatalytic performance by suppressing the serious recombination of photogenerated electron-hole pairs. Jincai Zhao et al. further confirmed that the role of Brønsted Acid on the surface of $SO_4^{2-}/TiO_2$ solid superacid is to photo-assist the cleavage of $\eta^2$ peroxide species($Ti-O_2$).

In recent years, some important progresses on the application of photocatalysis have been also achieved. A variety of advanced techniques, methods, and reactors have been developed to improve the efficiency of photocatalysis as well as to solve the key engineering issues. Meanwhile, the photocatalytic technologies have been widely used in atmospheric purification, indoor air purification, soil purification, water purification, industrial wastewater treatment, anti-pollution flashover high voltage insulator, self-cleaning coatings, anti-fogging glass, medical and health systems, and many other areas, greatly expanding the application of photocatalysis in daily life.

In conclusion, great progress on both the fundamental study and application research of photocatalysis has been achieved. This will lay a good foundation for the large-scale practical application of photocatalysis in environment & energy in the future. With the development of fundamental study, much deeper understanding of the key scientific issues on photocatalysis should be achieved, and a new photocatalysis reaction mechanism based on molecular-level should be set up for the design and synthesis of new photocatalysts with high performance. For practical applications, the efficiency of photocatalytic process should be enhanced with the development of advanced technologies, greatly accelerating the application of photocatalysis in environment & energy fields. We believe that the utilization of solar energy for photocatalytic water splitting and for artificial photosynthesis of organics in industry will come true in the future. If the large-scale application of photocatalysis in industry is achieved, large high-technical industrial chains based on photocatalytic technology will eventually form.

In theory,photocatalysis is considered as one of the ideal pathways to address the environment & energy issues, but it was not yet be fully realized. Most of the above-mentioned issues are on the way to be solved, and have not yet been fully addressed. As a researcher in photocatalysis, we should pay much attention to this promising and growing area and make it work better.

To close the talk in one sentence: the sun brings light and warmth to the world, the photocatalysis creates cleanliness and health to human.

Thanks for your attention!

**Prof. Xianzhi Fu** of the Chinese Academy of Engineering is the director of both National Engineering and Technological Research Center of Environmental Photocatalysis and State Key Laboratory of Photocatalysis on Energy and Environment at Fuzhou University, China. His main research interests concentrate on fundamental and applied photocatalysis, including design and preparation of photocatalyst, photocatalytic reaction mechanism, photocatalytic reaction kinetics and design of photocatalytic reactor. He has prepared a series of novel high-performance photocatalysts and developed many new technologies and methods to enhance photocatalytic efficiency, while industrializing several photocatalytic products and realizing the practical applications of photocatalytic technologies in such fields as environmental protection, building materials, power grid and so on. Thus far, he has published more than 300 peer-review papers in domestic and international journals, and obtained more than 40 patents authorized. The achievements have won a second prize of the National Science and Technology Progress Award, a first prize of the China People's Liberation Army Science and Technology Progress Award, and three first prize of Science and Technology Progress Award at provincial and ministerial levels.

# Biomaterials for Inducing Tissue Regeneration: The New Era of Biomaterials

**Xingdong Zhang**

National Engineering Research Center for Biomaterials, Sichuan University, Chengdu, Sichuan, China

**Abstract:** Biomaterials for inducing tissue regeneration refer to the biomaterials or implantable devices that are able to trigger required biological responses and subsequently regenerate and/or reconstruct damaged tissues or organs when implanted. They include biomaterials which can induce tissue regeneration by optimizing the design of the materials, tissue engineering products, as well as controlled release vehicles and systems of drugs and genes which enable tissue regeneration.

Traditional inanimate biomaterials had been proved very successful, but clinical practice has been demonstrating that they could not fully meet the clinical requirements, such as functionality, lifetime and so on. Therefore, the days of the traditional inanimate biomaterials have been passing, and the biomaterials science and its industry are undergoing a revolutionary change. Importantly, biomaterials that are able to induce tissue regeneration have evidently become a frontier and direction of biomaterials development, and will grow up into a key part of biomaterials industry in the near future.

The very key principle of the biomaterials to induce tissue regeneration is that inanimate materials could induce the regeneration of living tissues or organs. This philosophy, the evolution from nonliving to living, is a breakthrough to conventional wisdom. In this presentation, I will introduce our ground-breaking discovery, namely the regeneration of living bone tissues induced by nonliving biomaterials. Afterwards, the inducing mechanism will be discussed. Last but not least, the perspective of Tissue Inducing Biomaterials, i. e. the potential of non-osseous tissue inducing biomaterials, will be addressed.

**Prof. Xingdong Zhang** is a professor of Sichuan University, a member of the Chinese Academy of Engineering (CAE), a Foreign Associate of the National Academy of Engineering (U. S.), and the President of the Chinese Society for Biomaterials (CSBM). He is a Council Member of Tissue Engineering & Regenerative Medicine International Society-Asia Pacific Region. He is the director of both National Technical Committee on Dental Materials and Devices of Standardization Administration of China (SAC/TC 99) and National Technical Committee on Biological Evaluation on Medical Device of Standardization Administration of China (SAC/TC 248).

He has been engaged in biomaterials studies since 1983, especially on musculoskeletal medical therapies and biomaterial product development. He has received numerous awards, including National Chinese Natural Science Award (2007), Chinese National Science and Technology Progress Award (1998). Fellow of Biomaterials Science & Engineering, IUSBSE (2000) and Chinese National Expert with Outstanding Contribution (1992). He is the author of over 400 journal articles (~300 in English), the inventor of 22 Chinese patents, and has edited and co-edited 10 books. To date, he has obtained six Registration Certificates for Medical Devices issued by the State Food and Drug Administration of China (CFDA).

# Cornflour, Ketchup and Parts for Cars: A Review of Semi-Solid Processing

**Helen Valerie Atkinson**

The University of Leicester, Leicester, UK

**Abstract:** The thixotropic behaviour of semi-solid metals was first discovered by Flemings and his group at MIT in the early 1970s. When alloy was stirred as it solidified, and then reheated into the semi-solid state, it behaved in a surprising way; on shearing it flowed with the consistency of heavy machine oil but when allowed to stand it could be handled like a solid. This behaviour (which is very like that of tomato ketchup) was then exploited in a family of semi-solid processing techniques, including: thixoforming, thixoforging, thixocasting, rheocasting and rheoforming. "Thixomolding" (based on injection moulding) is used by numerous companies, particularly in Japan and the US to produce magnesium alloy components e. g. for portable computers and cameras, but it is not suitable for aluminium alloys. Aluminium alloys have been widely produced for automotive applications, mainly by thixocasting and thixoforming-type processes.

In the last few years, there has been a great increase in interest in semi-solid processing in China, India and other countries with many new ideas under investigation. This lecture will review the current situation.

**Dr. Helen Valerie Atkinson** has a first class degree from the University of Cambridge and a Ph. D. from Imperial College. She is a Fellow of the Royal Academy of Engineering (the highest honour for an engineer in the UK) and a Vice President. She was made a Commander of the British Empire (CBE) for national services to engineering and education in the Queens New Year' s Honours in 2014.Professor Atkinson is Head of the Department of Engineering at the University of Leicester and leads the Mechanics of Materials Research Group. She has an honorary Ph. D. from the University of Liege in Belgium,and in 2011 received the Lee Hsun Lecture Award from the Institute of Metals Research,the Chinese Academy of Sciences and the Shenyang National Laboratory for Materials Science for her " outstanding contribution in the field of materials science and engineering". She has been a Visiting Professor at the Arts et Metier ParisTech (one of the French Grand Ecoles premier universities) since 2010 and was made a Visiting Professor at the General Research Institute for Non-Ferrous Metals in Beijing in 2013. She was elected the first woman President of the Engineering Professors' Council (the body which represent engineering throughout higher education in the UK) and has been given a national award as a Woman of Outstanding Achievement in science, engineering and technology for "leadership and inspiration to others".

# The Future Product/Process Development Challenges in Chemical and Material and Allied Engineering Fields

**K. V. Raghavan**

Indian National Academy of Engineering, Indian Institute of Chemical Technology, Hyderabad, Telangana, India

**Abstract:** Fundamentally, Chemical and material engineering have derived their advanced scientific inputs from quantum and solid state chemistry, electronic physics, applied mechanics, mathematics and biology. Their future product/process innovations greatly depend on molecular level understanding of their transformations over hierarchically organised micro or nanotime and length scales coupled with deeper insights into the science of their formation chemistry. Multiscale modelling of molecular motions, manipulation of individual or group of molecules to create new microstructures with higher reactivity and preproduction tailoring of new molecular entities and their properties provide exciting opportunities in new product design and development in these sectors.

Engineering understanding of microbial living systems is very vital for designing and forming chemically definable biomolecular entities of consumer utility as well as those possessing high biological activity on human or animal systems. Similar approaches are attracting the attention for discovery and development of new biomaterial entities. Explaining complex biotransformations in terms of well accepted engineering concepts continues to be a challenge in evolving new biorefining methodologies to provide a wide range of ecofriendly chemical products to the petrochemical and speciality chemical fields.

From cleaner technology considerations, the development of self optimizing chemical systems to produce a single desired product is an exciting option for the development of novel therapeutics catalytic systems and ultrapure specialities. Catalysis has a strong interphase with chemical reaction engineering to provide a wide variety of process technologies in chemical and allied sectors. The recent efforts to tailor the microenvironment around heterogeneous catalytic entities, simulation of molecular interactions within catalyst pores, computational fluid dynamic (CFD) modelling of reactor internal hydrodynamics and allied initiatives have contributed to the achievement of higher specificity and reaction rates in commercial chemical process systems.

An attempt is made in this presentation to demonstrate how the aforementioned and other engineering advances have changed the complexion of chemical and material fields in terms of predicting and improving their performance under the influence of process constraints typically encountered in commercial manufacture.

My colleagues and I are honored to attend today´s seminar. We are also very happy to see so many young scholars. The speeches on biomaterials and processing technique just presented showed many new research fields and relevant progress. I will introduce future products, technology, research and challenge of chemical and materials engineering, and how to provide a good environment for future research and development.

Speaking of chemistry and materials, it refers to both chemical and material science. If you want to study biological material, you will have to study biological sciences. When we talk about the process of product development, the development process should be a clean process. This is a topic we are now keen to discuss. Chemical engineering is now becoming more and more multi-disciplinary and interdisciplinary, especially considered the expanding research contents. This is different from the 1950s or 1960s, when chemical engineering were totally different from today. Currently, it is necessary to get a comprehensive, through, and interdisciplinary understanding of the entire process.

The research contents are rapidly increasing, including vibrancy in molecular design. It also shows a more and more interdisciplinary trend. It is important to have innovative technology and development of new products. Meanwhile, for science and engineering, everything has some core knowledge and technology as well as adjacent disciplines, such as chemical, physical and biological sciences. Once you have the basic knowledge, you can conduct interdisciplinary research and affect the whole process.

A lot of research has been done in chemical engineering. It is a multifaceted process. A lot of research is at the micro level, especially when the process needs to be involved with some of the new knowledge. Flow dynamics, including equipment design will also be applied to modern chemical engineering science and technology. Modern chemical engineering can be applied from the macro level to the nanometer level.

Boundaries between chemical and material engineering are fuzzy at best. Some people treat chemical engineering as applied organic chemists and material scientists as applied condensed matter physicists; but it is not totally true. Chemical engineering actually provides opportunity galore to material science in discovery, design, development and manufacture. Typical examples include graphene nano-composites and other composite materials. The chemical molecular tools we discuss today are for material synthesis, including corresponding nanotechnology, which enables us to better study gene and help medical develop in corresponding field.

Now there are some molecular models which can be very precise. This achievement relies on the development of nuclear chemical engineering. Biology is incredibly nonlinear and less predictive as compared with chemistry. Actually, a large number of chemical processes are also biological processes. For instance, here we have Michaelis-Menten kinetics of enzymatic reactions, which can be expressed by ODEs. Feedback and feedforward control logic can be used for biological systems with appropriate modifications. Recent engineering exploration of genetic, genomic and proteomics fields, however, shows the difficulty of fully comprehending their dynamics. Now with progress in biological sciences, the total equation can be better studied. Moreover, application in gene is rapidly developing with some recent progress in engineering.

Engineering living systems on chips have drawn attention all over the world. These systems can help us have more discoveries in medicine. Models can be developed to reproduce dynamic properties of human cells, tissues and organs. Functions of lungs, blood vessels, kidney, bones and brain are successfully simulated and chips engineering study can be conducted on living system. Currently, biotechnology development becomes more and more complex as well as more achieved.

Molecular level understanding of structure-activity relationship is vital for new biologically active chemical and material engineering. In fact there are some commonness among these three areas, including new molecules found in medical field, new chemical catalyst field and new materials. A relationship can be seen between new chemical entity and human body receiver. Human body environment needs to be simulated. But for catalyst, it is a micro medium; the specific environment of catalyst generated needs a very precise measurement. Reaction environment needs to be established between M1 and M2. Combination library size is more than 10 000; yet practice is quite limited.

Here are some examples. The first one is molecular modeling. It is important we know their activities and select molecules to be tested. Also we need to know what kind of energy will result in molecular changes, and lead to diseases. Information of chemical and biological structures is needed to decide whether this molecular modeling will work. This is the balance of synthesis. We will have hundreds of candidates, so we need to choose the one we need. There is a micro chemical plant. Here we have raw material going to the plant and final products. This typical reactor can synthesize more than 100 kinds of materials in different processes.

An experiment about six years ago revealed that there was a linear process. Energy was required to realize this process. It plays a role in the systems. The reaction of the catalyst was as important as what had happened during the process. The novel concepts of environmentally clean chemical naturally came out. It contained reactive separations, designing catalysts environment, external stimulation and using self regulation molecules. The concepts of atom economy by Professor H Trost in 1991 and E-factor by Dr. Anastas and Dr. Warner in 1998 laid the basic foundation for clean chemical processing.

Another important technology is cross disciplinary catalysis. Multi-functional and cross disciplinary research is very complex. Researchers in varied disciplines need to work together. I'd like to show you an interesting development about extraction, distillation and absorption. The process meeting the requirements of the objective needs to be chosen. It is hard to model the system, but it can become more specific. This is a complete process of distillation, absorption and second-time absorption. Many such examples can be found in the field of chemical industry. The next task is to create a conductive microenvironment around a catalyst particle. Positioning of desired reactants around active sites is crucial for achieving effectiveness and specificity in binding. Available options for creating microenvironment in catalytic systems include heterogenization of a homogeneous catalyst, creation of a new phase, creation of microcages or microcavities or micro structures, tailoring morphology of liquid-liquid dispersions and encapsulating metal complexes in molecular sieves.

This is solid acid catalyzed nitration. One of the research goals is to realize the maximization of the utility of catalyst. In the reaction distillation, it is necessary to keep environment friendly, to choose the right catalyst, and to realize its economy. The multiphase mass transfer in vapor-liquid-liquid-solid system includes organic dispensed, vapor space and liquid film. Below the liquid film is catalyst particle. When the catalyst is filtered, pressure is enhanced and thus the process starts. This is a process of vapor into liquid and liquid into solid. Novel reaction media can be adopted for higher process efficiency, such as supercritical fluids, pseudo liquids and micro emulsions.

Additional driving forces such as microwaves and ultrasound can be used to speed up reactions. The membrane here can achieve goals by permeation. Conventional method takes 9 hours, 9 kW · h and the productivity is only 73%. The novel technology requires only 0. 01 kW · h and the productivity reaches more than 80%. The difference is obvious. Future is for self optimizing or regulating chemical systems to produce a single product without any impurities. This will open a new chapter in clean chemical processing.

Self organization is an interesting phenomenon. It arises out of local interactions between components of disorderly system to become orderly. It happens in physics, chemistry, biology, networks, human society, psychology, algorithms etc. It is a key concept in supramolecular chemistry.

Modeling of such systems is very interesting. It is a challenge from unconventional engineering. The first one is reaction-diffusion model. It has some weakness; then here comes the cellar automata model. The basic feature is grid sites interacting with neighbors. It exhibits desired patterns, but cannot simulate irregular networks. The third one is agent-based model. Agent interacts in complex and dynamic ways and it simulates social networks well. Analytical, reference design and evolutionary algorithm based design approaches are employed in chemical engineering.

Self optimizing system concept is solving complex production problems. Handling complex products as well as innumerable customer demands is highly challenging. Self optimizing manufacturing systems with changing boundary conditions offer unique solutions. They cover machines, products, logistics, man-machine interfaces, work places and supplier integration architectures. Attention needs to be paid on multiple disciplines whose development is of great significance.

**Dr. K. V. Raghavan** is a fellow of the National Academy of Engineering, Indian Institute of Chemical Engineers (IIChE) and A. P. Academy of Sciences and a distinguished fellow of University of Grants Commission (UGC). He did his B. Tech from Osmania University in 1964; M. S. and Ph. D. from the Indian Institute of Technology (IIT), Madras. He joined CSIR service in 1964 and worked in three national laboratories. He was appointed as the Director of Central Leather Research Institute (CLRI), Chennai in 1994. He took over the Directorship of Indian Institute of Chemical Technology, Hyderabad in 1996. On successful completion of this tenure, he was appointed as the Chairman of Recruitment and Assessment Centre of DRDO, Ministry of Defence, Government of India in May 2004. Dr. Raghavan has taken up INAE Distinguished Professorship in October 2008 at IICT, Hyderabad. His research group is engaged in reaction engineering studies pertaining to green processes. Chemical process development and design, reaction engineering, simulation and modeling and chemical hazard analysis are his areas of specialization. His basic research contributions cover simulation of complex reactions in fixed bed reactors, hydrodynamics of multiphase reaction systems, envirocatalysis for clean processing, zeolite catalysis for macromolecules, thermochemistry and kinetics of charge transfer polymerization and modelling of chemical accidents. His current research activities are in process intensification of water gas shift reaction, catalytic $CO_2$ decomposition, analysis of $CO_2$ capture technologies and characterization of the reactivity of ionic liquids.

# Energy and Mineral Resource Development and Utilization: Past, Present and Future

**Z. Xu, M. Wyman, J. Masliyah, K. Cadien**

Department of Chemical and Materials Engineering, University of Alberta, Edmonton, Alberta, Canada

**Abstract**: Civilization has always depended on energy and mineral resources. Ever since the beginning of the Stone age millions of years ago, when mankind used sharp edges and points for self-defense and for hunting for food, advancements in civilization have been closely linked with the development of natural resources processing and utilization through extraction and/or fabrication. Development continued through the Bronze age and Iron age, leading to development of glass, ceramics and modern metals such as steels/stainless steels and alloys/super alloys, polymers, diamond, composites, and silicon which revolutionized current telecommunications and brought us to the information era. By mimicking nature with nanotechnology, life science is making sharp advances. All our advancements have heavily relied on the use of natural mineral resources and fossil fuels. Continuing increases in the world's population and demand for better lives have put tremendous pressure on the utilization of non-renewable mineral and energy resources. Although recycling of metals could partially alleviate the demand for metals from natural resources, it does not supply the needs of increasing populations and trend of improving life styles. For fossil fuels, the challenges are even bigger due to their nonrenewable/nonrecyclable nature. As a result, we are forced to develop increasingly lower grade mineral resources while searching for alternative fossil fuels.

In this paper, we will provide a historical overview of the critical role the development and utilization of mineral resources have had in the advancement of civilization and the impact of the extensive energy utilization that was required for extraction and fabrication.

With the rapid depletion of easily-processed natural resources, processing of lower grade and more complex mineral resources becomes inevitable, which requires even higher energy consumption and presents more severe threats to our environment. Increasingly high energy utilization per capita, coupled with ever increasing population calls for alternative energy resources, in particular the harvesting of solar energy, which again requires development of revolutionary materials. Meanwhile, unconventional energy is becoming increasingly critical to bridge the gap between the current fossil fuel based energy sources and the effective harvesting of renewable energy such as solar energy. An example will be given, namely oil sands development as an alternative fossil fuel source to secure the energy need before renewable energy sources become economic and competitive for large scale demands.

## 1 Introduction

Materials have been a key enabler of the growth and advancement of the human race far back into prehistoric times, as the evidence for tool fabrication long predates written history. Today there is saying amongst material scientists that "*without materials there is no engineering*" that has a ring of truth to it. There is evidence that humans have been using wood for spears and stone tools for many millennia. A schematic diagram shown in Fig. 1 indicates the approximate time at which new materials were widely

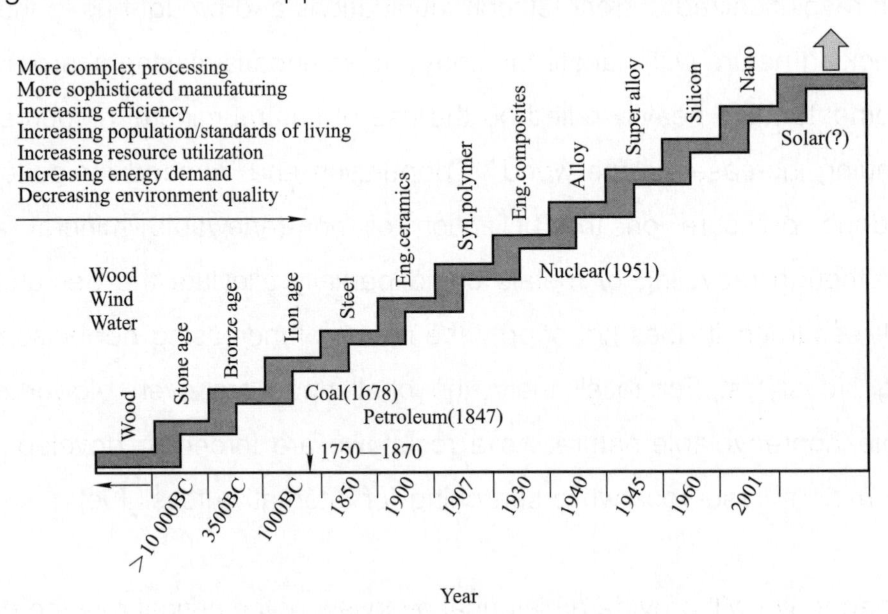


**Fig.1 Schematic staircase diagram indicating the approximate time at which new materials were widely adopted**

adopted, which coincides with major advances in human civilization. It should be noted that as new materials became widely adopted the older materials continued to be used, often in more effective and innovative ways. For example, today we still use wood for building structures and houses, and making furniture.

## 2 The past

Twelve thousand years ago when farming started, materials such as wood and stone had already been in use for millions of years as tools and weapons. Fabrication of stone and wood into tools and weapons required skill, but the technology was fairly straightforward. Starting in the Bronze Age ( starting ~ 3500 BC ) more complex processing was required. Fabricating bronze requires the invention and development of the technology for smelting copper and alloying it with tin. Of course different societies around the world entered the Bronze Age at different times, depending on the availability of ores, or the ability to trade for bronze from areas that had bronze technology. In some parts of the world the Iron Age ( starting ~1000 BC ) overlapped with the Bronze Age, but generally the Iron Age is considered to be the time that followed the Bronze Age, in which the use of iron was prevalent.

Processing of iron ore required temperatures 1200℃ and the development of the Bloomery furnace. Early iron objects were made from meteoric iron. One of the main driving forces for both the Bronze and Iron Ages was the development of better swords and spears, that is, weapons for hunting and war for the feeding and protection of an increasing population as shown in Fig.2. These two ages lasted for an average of over 2500 years each, but starting in the early $18^{th}$ century with the discovery of the use of coke to produce cast iron and in the $19^{th}$ century with the discovery of petroleum, the pace of innovation accelerated.

The acceleration in technology innovation started in the late $18^{th}$ century and can be traced to the fact that many of the materials developed since then require increasing amounts of energy to produce them. The development of these materials was aided by the development of cheap and high density sources of energy such as coal and petroleum. This was combined with a renewed interest in the fundamental understanding of how and why things worked. The invention of the microscope led to the discovery of microstructure in metals, which in turn led to improvements in cast iron and an understanding of the role of carbon in iron. The invention of steel by using oxygen to

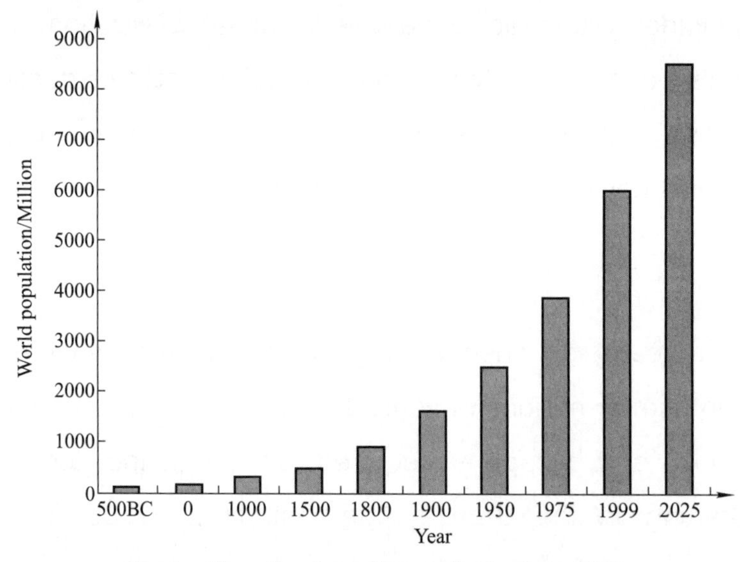


**Fig.2 Growth of world population into 2025**

eliminate carbon was a major breakthrough that added to the industrial revolution starting in the mid $19^{th}$ century. Materials enabled the harnessing of the power of steam, which started a revolution in infrastructure with the development of railroads, engines and factories powered by steam. The generation of steam requires a source of energy (heat) which was accomplished by burning wood or coal, leading to increased pollution. The discovery of vast supplies of oil at the end of the $19^{th}$ century led to the development of an energy source that had twice the energy density of wood or coal as shown in Table 1. In addition, since oil is a liquid it is easier to extract than coal, and it takes up a much smaller volume than the equivalent amount of coal. Furthermore, the liquid fuels are easier to store and transport, and to use in systems that need high intensity energy. Discovery and utilization of oil led to the development and proliferation of the internal combustion engine, which in turn led to the development of the automobile, busses, aircraft, and diesel-electric trains, for example. The internal combustion engine also led to the development of improved metal fabrication technologies, new alloys and engineering ceramics. Oil also led to the invention and development of polymers.

The last sixty years has been driven by the invention of semiconductor technology, specifically the invention of the silicon transistor and the integrated circuit, and we call this time period the silicon age. Today integrated circuits are ubiquitous in every field, and enable the connection between all the peoples on earth. Semiconductor sales total ~ 330 US MYMB/a which enables a global electronics market of more than 3 US MYMT. The electronics market (2012) for the top four countries is shown in Fig. 3. The foundation of this market is not based solely on the invention of the transistor and the integrated circuit, but also on the discovery of a breakthrough technology for the purification of semiconductors (and metals) by Pfann at Bell Labs.

**Table 1 The specific energy of different fuels**

| Fuel | Specific energy /(MJ/kg) |
|---|---|
| Diesel/Fuel oil | 48 |
| Propane/Butane | 46.4 |
| Gasoline (petrol) | 44.4 |
| Coal | 24 |
| Wood | 16.2 |

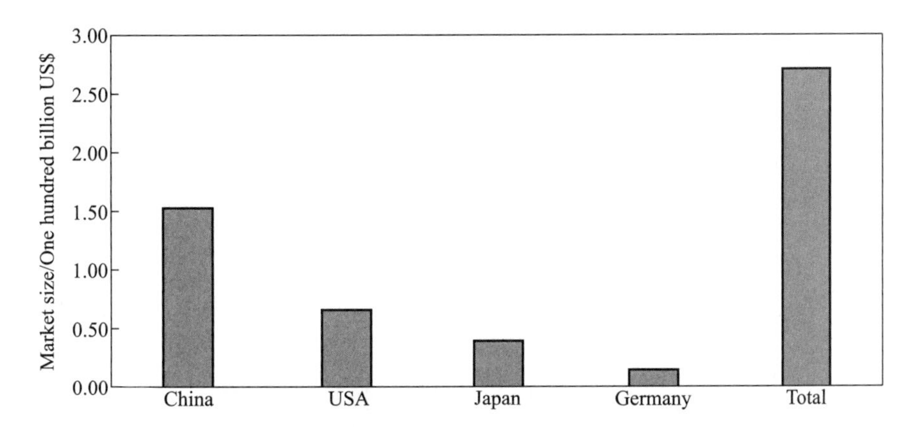


**Fig.3 The size of the electronics market in 2012 for China, USA, Japan and Germany**

The invention of the transistor coincided with the increasing awareness of the impact of humans and industry on the environment. In fact Rachel Carson's seminal book on the environment, *Silent Spring*, was published in 1962, two years before the invention of the planar integrated circuit. Since this time the concern for the environment has led to renewed interest in more efficient fabrication methods, and the application of novel materials to address environmental concerns such as energy consumption and global warming.

## 3 The current

As we move into the future, mankind continues to strive for better lives while the population continuing to grow at an exponential rate as shown in Fig.2. The demand for materials derived from natural resources will see a corresponding increase, while nonrenewalble natural resources become increasingly depleted. To meet the ever increasing need for materials, mankind has been forced to exploit lower grade, more challenging natural resources, which requires higher energy intensity to develop. Although effective recycling of materials will alleviate the pressure of increasing demand of materials to some degree, however recycling of materials also requires the use of non-renewable energy resources. Despite great efforts to develop and implement renewable energy resources such as solar, wind, hydro, geothermal and biomass, we will continue to see a greater rate increase in demand for non-renewable fossil fuels well into 2040, as shown in Fig.4. This is due to the anticipated increase in population (Fig.2) and increasing living standards of large population countries such as China and India.

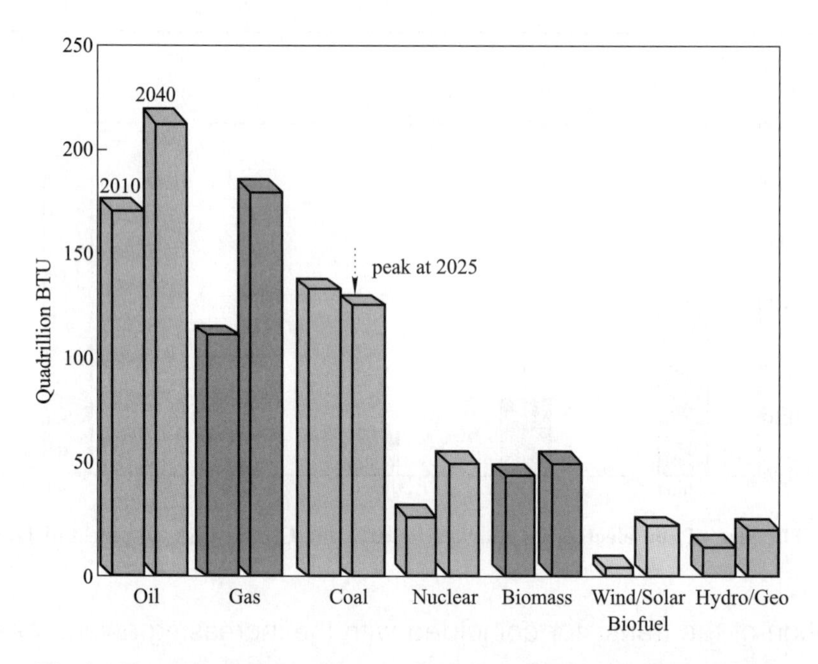


**Fig.4 The outlook for energy: A view to 2040 (ExxonMobil, 2014)**

Such an increase in non-renewable energy demand well into the foreseeable future has motivated the oil industry to search for unconventional oil resources. Rapid development of technologies for exploiting Canada's vast oil sands resources and fracturing of shale to realise hitherto unreachable oil reserves in the United States of America serves as good examples in this strategic direction. Unfortunately, exploiting these types of energy resources requires higher energy intensity operations which cause more serious environmental concerns due to air and water pollution while simultaneously increasing the pressure on fresh water supplies and threatening the natural ecology.

Taking oil production from Canadian oil sands by the mining-extraction method as an example, it takes about 20% of the energy to produce the marketable fuels, while consuming approximately three times the volume of water. Such a high energy requirement is linked to the complex processes of oil production from oil sands as shown schematically in Fig.5. In this method as schematically shown in Fig.5 (a), the mined oil sands ore is crushed prior to making slurries with chemical additives and hot water. The hot water is necessary due to excessively high viscosity of bitumen which does not flow at ambient temperature. The slurry is transported by hydrotransport slurry pipelines where oil sands are digested to liberate bitumen from sand grains, which is considered an innovation that enabled the oil sands industry. During the slurry hydrotransport, the liberated bitumen droplets are attached to entrained air, forming the aerated bitumen of density much lower than the density of processing medium-water. After a few kilometers of hydrotransport, the conditioned slurry is then discharged to a stationary separator, known as the primary separation cell where the aerated bitumen floats to the top of the slurry as bitumen froth typically containing 60 wt% bitumen, 30 wt% water and 10 wt% solids after removal of air. To further remove the solids and water, organic solvent, typically the by-product of bitumen upgrading is added to the bitumen froth to dissolve the bitumen, increasing the density difference between organic and aqueous phases for effective phase separation, again under the gravity force using inclined plate settlers, hycrocyclones and/or centrifuges, collectively known as froth treatment. The organic phase is then sent to solvent recovery unit to recover the solvent for recycle, while producing bitumen containing typically less than 1 wt% water and solids, suitable for downstream upgrading as schematically shown in Fig.5 (b). The aqueous phase from both the primary separation cell and the froth treatment are discharged into large volume tailings ponds where solids are separated from water for recycle. Such process could comfortably lead to a bitumen recovery greater than 90%, producing 1 barrel bitumen from around two tons of oil sands containing around 9 wt% bitumen.

Bitumen from the extraction process is upgraded by either coking or thermal cracking, before going through hydrocracking to produce synthetic sweet crude. The coking and thermal cracking are mainly to reduce the size of molecules, while hydrotreating is mainly to remove sulfur and nitrogen from the crude to produce product suitable for refining. During the upgrading, significant hydrogen is needed, which is produced from natural gas, while waste heat generated could be used in extraction process.

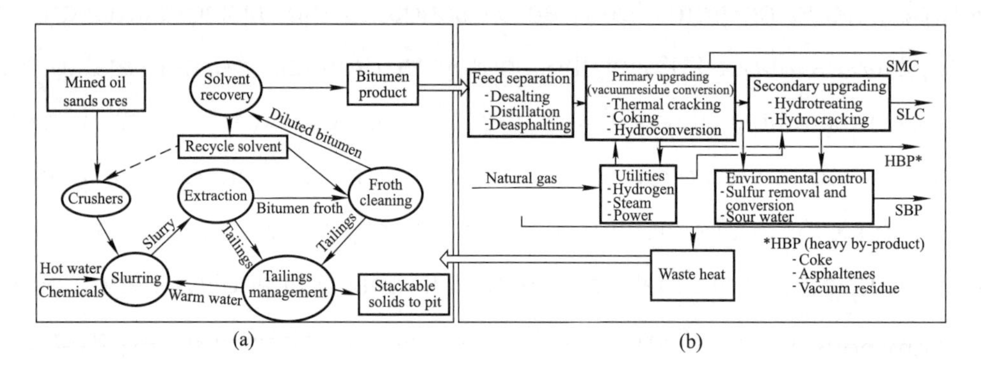


**Fig.5 Schematic process diagram of oil production from Canadian oil sands using mining-extraction (a) and bitumen upgrading (b) processes**

It is evident that much effort is needed to produce oil from oil sands, not only requiring more energy and involving complex operations, but also producing waste tailings and heavy by-product, along with greenhouse gas emissions. Similar environmental challenges exist for shale oil fracturing. Unfortunately to fill the gap between now and the future when the technology is available to generate sufficient clean energy for mankind, we are forced to use fossil fuels as shown in Fig.4. While we are developing new materials and technologies for the harvest of renewable energies, we are committed to developing technologies for more responsible use of fossil fuels with the aim of minimizing the environmental impact of fossil fuel production and utilization. One approach is to reduce the operating temperature of the extraction process. A simple energy balance analysis shows that a reduction in the operating temperature by 1℃ reduces the energy to produce one barrel of oil by 10 MJ. With a current production rate of 800 000 barrels of bitumen per day by mining-extraction method, a reduction of 5208 $m^3$ of gasoline could be easily realized if one could reduce the operating temperature of the process from the current 45℃ to ambient 20℃. Such a reduction in energy would not only reduce the operating cost, it also would also reduce the greenhouse gas emissions by 13 000 tons $CO_2$(eq) per day.

A close look at the current operations in Fig.5 (a) shows that thermal energy (hot water) is used to reduce the viscosity of bitumen so that it flows and can be liberated. An alternative method to reduce bitumen viscosity is to add the solvent to bitumen, as practiced in froth treatment. As shown in Fig.5(a), solvent is used downstream in froth treatment. An idea is to add a portion of this solvent upfront as indicated by black dash arrow in Fig.5(a) so that the bitumen viscosity could be reduced to a value similar to that created by hot water addition. Such a replacement would eliminate the use of hot water. The new process known as aqueous-nonaqueous hybrid extraction process shown by the schematics with dashed arrow in Fig.5(a) has been proven in laboratory scale experiments to be feasible, leading to a process that does not need other chemical additives in the process. Similar energy savings could be realized by implementing gasification of petroleum coke produced during bitumen upgrading. Such an integrated energy system could reduce natural gas consumption while producing a stream of high $CO_2$ concentration for $CO_2$ storage or enhanced oil recovery. This oil sands operation example gives us good guidance on how to reduce the environmental impact of utilizing fossil fuels to meet our energy needs in the next decades. More importantly, we could not over-emphasize the critical importance of fundamental research in assisting revolutionary technology development, if we were to succeed security of supplying energy and natural resources whether it is renewable or non-renewable.

## 4 The future

Materials designated as enablers of a human epoch are done so in retrospect. In this section we will not attempt to predict the future, but rather we will review some of the trends that will continue on into the future. In the last decade, discoveries that were made more than 40 years ago have been brought to the forefront due to concerns about the environment, and the desire for alternate renewable energy sources as well as more efficient use of resources as shown in Fig.6.

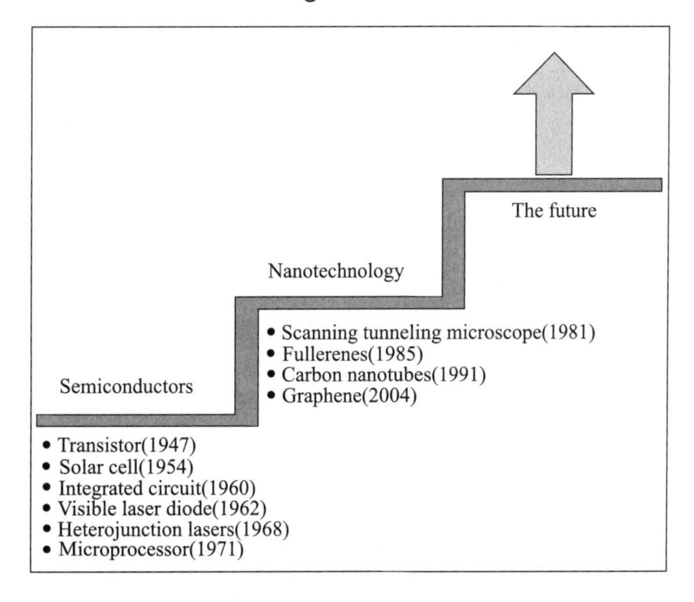


**Fig.6 Staircase schematic plot of innovations and discoveries during the last 60 years**

The two technologies are light emitting diodes (LEDs) and solar cells, invented in 1962 and 1954, respectively. Both technologies required almost 50 years to make a significant impact on humanity. Discoveries made in the last twenty years, namely carbon nanotubes and graphene, have led some to predict the age of carbon being the future materials epoch. However, the jury is still out on these materials and the future is not clear. There is evidence that progress in biotechnology and nanobiology may lead to these materials having increased prominence. However, it is clear from the viewpoint of 2014 that the trend is for the need for more efficient use of energy sources, more efficient devices, new renewable energy sources, more efficient internal combustion engines, and recycling of more materials, etc. will all continue to grow, and possibly lead to a self-sustaining future.

## References


English Heritage. Introduction to heritage assets: Pre-industrial ironworks. https://www. english-heritage. org. uk/publications/iha-preindustrial-ironworks/preindustrialironworks.pdf.

Gosselin P, Hrudey S, Naeth M, et al.2010.Report of The Royal Society of Canada Expert Panel: Environmental and Health Impacts of Canada's Oil Sands Industry.Royal Society of Canada, Toronto.

Harjai S, Flury C, Masliyah J, et al. 2012. Robust aqueous-nonaqueous hybrid process for bitumen extraction from mineable Athabasca Oil Sands.Energy Fuels, 26: 2920–2927.

http://www.semiconductors.org/industry_statistics/global_sales_report/.

http://www.statista.com/.

Masliyah J, Czarnecki J, Xu Z.2011.Handbook on theory and practice of bitumen recovery from Athabasca Oil Sands. Vol.1: Theoretical Basis.Kingsley, Calgary.

Michael F Ashby.2011.Materials selection in mechanical design.4th Edition.Burlington M A: Elsevier.

National Geographic. The Genographic Project. The Development of Agriculture. https://genographic.nationalgeographic.com/development-of-agriculture/.

Pfann W G.1952.Principles of zone melting.Trans.American Institute of Mining and Metallurgical Engineers, 194: 747–753.

Smithsonian National Museum of Natural History.http://humanorigins.si.edu/evidence/behavior/tools.

The outlook for energy: A view to 2040 (2014, ExxonMobil).http://cdn.exxonmobil.com/~/media/Reports/Outlook%20For%20Energy/2014/2014-Outlook-for-Energy.pdf.

**Prof. Z. Xu (Zhenghe Xu)** graduated with B. Sc. and M. Sc. degrees in Minerals Engineering in 1982 and 1985, respectively, from Central-South Institute of Mining and Metallurgy, Changsha, China; and Ph. D. in Materials Engineering in 1990 from Virginia Polytechnic Institute and State University, Blacksburg, Virginia. He worked as a Research Associate at the Virginia Center for Coal and Mineral Processing and then as a postdoctoral fellow at the University of California, Santa Barbara (Chemical and Nuclear Engineering) until 1992. He was appointed as an assistant professor in the Metallurgical Engineering at McGill University in September 1992, and moved to the Department of Chemical and Materials Engineering at the University of Alberta as an associate professor in January 1997. After promoted to professor in 2000, he was appointed as NSERC-EPCOR-AERI Industry Research Chair in Advanced Coal Cleaning and Combustion Technology from 2002 to 2007, Canada Research Chair in Mineral Processing in 2007, and NSERC-Industry Research Chai in Oil Sands Engineering in 2008. Dr. Xu's main research area is interfacial sciences as applied to natural resources processing and utilization. He published close to 300 peer-reviewed scientific journal papers and 57 technical conference proceeding papers along with three US patents and one Canadian patent. He co-authored one book, co-edited two books and co-authored ten book chapters. Dr. Xu was named Teck professor in 2007, elected to Fellow of Canadian Academy of Engineering in 2008 and Fellow of CIM in 2010, and awarded APEGA Frank Spragins award in 2012 and The Teck Environmental award in 2013. He is currently 3rd Vice President of Metallurgical Society of CIM, Canada.

# Biohydrometallurgy: Biotech Key to Unlock Mineral Resources Value

**Guanzhou Qiu, Weimin Zeng**

School of Minerals Processing and Bioengineering, Central South University, Changsha, Hunan, China

Key Laboratory of Biometallurgy, Ministry of Education, Changsha, Hunan, China

**Abstract**: At present the global high grade mineral resources decreased in a large scale, and the depleted phenomenon of mineral resources is increasingly serious. The traditional treating methods like flotation and smelting can't deal with low-grade mineral resources efficiently, however, biohydroemtallurgy have been proved to be a feasible and effective technology in the treatment of the complex, difficult to handle, low grade mineral resources. Because there are four diversity in the biohydrometallurgy process, such as mineral species diversity, microbial community diversity, diversity of biochemical reaction and the complex diversity of leaching process parameters, block the clarify of microbial leaching behavior. This would hinder the development of the bioleaching theory. Aiming at the above problems, the bioleaching innovation team in Central South University in China has developed suitable technology or skill for the research of microbial community genome, functional genome and metagenomic chip. The theoretical research results achieved from macroscopic to microcosmic, from qualitative to quantitative, and from phenomenon to essence of bioleaching. These clarify the structure of the microbial population, community dynamics and function during bioleaching, reveals the interface reaction mechanism among mineral surface-microorganism-solution, lay the groundwork for the explanation of bioleaching mechanism. These theories achievements have been applied in the bio-mining industry, such as copper, gold, uranium resources, and obtained high economic and social benefits. This would provide the abundant metal resources to guarantee the sustainable development of the national economy. As a result, the workers and researchers in the world consistently advocated to establish the international society of biohydroemtallurgy, aims to " use the biotech key to unlock the door of mineral resources value" .

## 1 Introduction

The earliest record of biohydrometallury in China was described in the book titled *ShanHaiChing* in the $6^{th}$ or $7^{th}$ century BC, and that focused on " in SongGuo Mountain, LuoShui River flowing out and into WeiShui River, which contain much copper" . In Han Dynasty, HuaiNan King Liu An had written a book titled *Huai Nan Wan Bi Shu* (in the $2^{nd}$ century BC), and a record about copper extraction from acid mine drainage (AMD): " copper was obtained when iron was put into BaiQing solution. " In Tang and Song Dynasty (in the 600–960 years AD), copper hydrometallurgy factories were established and the annual highest copper production reached more than 500 000 kg.

In the 1950s when the international biohydrometallurgy research sprang up, Professor He Fuxu in Central-South Institute of Mining and Metallurgy (now called Central South University) founded a biohydrometallurgy laboratory in 1958, and began to carry out bioleaching research in series. In 1960, Microbial Institute of Chinese Academy of Sciences began the biohydrometallurgy industry experiment in Tongguanshan Copper Mine. In 1970, they reached a heap bioleaching amount of 700 tons of low grade uranium ore. In 1995, Central South University of Technology (now called Central South University) cooperated with Dexing Copper Mine in Jiangxi Province to exploit the low grade copper waste ore and recycle copper. In 1997, a biohydrometallurgy factory with annual production of 2000 tons cathode copper was established. In 1999, Central South University cooperated with Oak Ridge National Laboratory in USA to carry out the ecological and genomic research of bioleaching microorganisms. In 2000, China's first plant for bio-pretreatment of refractory gold ore with annual treatment of 50 tons gold concentrate was officially launched. In 2005, a bioextraction plant with a capacity of 30 000 tons of cathode copper was built in Zijin Mining Company, and the cathode copper purity reached the international level A standard.

Currently, Chinese government has attached a great importance on biohydrometallurgy, and is still now carrying out integrated applications of bioleaching technology in copper, gold and uranium extraction to ensure the China's economic reserves of strategic mineral resources.

In terms of copper resources, China is now the top copper consumer, but its present copper production from chinese ores accounts for only about 20% of China's copper demand. Biohydrometallurgy technology permits to exploit the low grade copper resources in China, and can be also used in foreign low grade resources, to increase the copper production. In terms of gold resources, the present refractory gold mineral resource accounts for about two-thirds of the gold reserves, the development and utilization of these resources are very difficult, and biohydrometallurgy pretreatment technology can effectively overcome gold mineral treatment difficulties. With the promotion of biohydrometallurgy technology, China has become the first producer in the world for four consecutive years. The application of this technology has vital significance to keep China as the first gold producer, and internationally competitive.

In terms of uranium resources, biohydrometallurgy technology can enable an efficient use of a large number of idle or abandoned uranium sulfide resources in China, and is expected to make exploitation grade of uranium resources decrease from the current one over one thousand to three over ten thousand, and thus greatly increase the economic mining exploitation of uranium deposit. Currently in China, the main biohydrometallurgy research units are: Central South University, Beijing Non-Ferrous Metal Research Institute, Institute of Process Engineering and Institute of Microbiology in Chinese Academy of Sciences, Shandong University, Changchun Gold Research Institute, Beijing Research Institute of Chemical Engineering and Metallurgy of China National Nuclear Corporation and Kunming University of Science and Technology, etc. The researches carried out by the above units mainly focus on three aspects: ① microbiology of bioleaching; ② microbial-mineral interaction research; ③ multi-factor strong correlation of bioleaching system.

In recent years, the Chinese government has given strong support and funding on biohydrometallurgy, and has established a number of national science and technology plans or projects, including "National Basic Research Program of China" (973 Program)," National High Technology Research and Development Program" (863 Program)," National High Technology Industrialization Demonstration Project", etc. National Natural Science Foundation also gave a lot of project funding, only in its Engineering and Materials Department, the number of approved projects increased from 10 in 2000 to 50 in 2010, and the funds increased by 10 times. Current research has made great achievements in China Nonferrous Metal Mining (Group) Co.,LTD,China National Gold Group Corp.,China National Nuclear Corp.,Jiangxi Copper Corp. and Fujian Zijin Corp. and other companies to achieve industrial applications.

## 2 Progress of biohydrometallurgy in China

### 2.1 Macroscopic to microscopic views of biohydrometallurgy

In China,the long-term research about biohydrometallurgy only stood at macro level. Metallurgists usually selected the AMD (Acid Mine Drainage) for bioleaching by color (Fig.1),but couldn't know what was the microorganisms composition,what were their role and how they worked. On the other hand,microbiologists only paid attention to the selection of strong activity strains (Fig.3) by isolation of pure culture or microscopic detection (Fig.2) and put them into bioleaching system,but they are not certain about the microorganisms growth trends,functions and effects into the bioleaching system.

**Fig.1 The deep red acid mine drainage (AMD)**

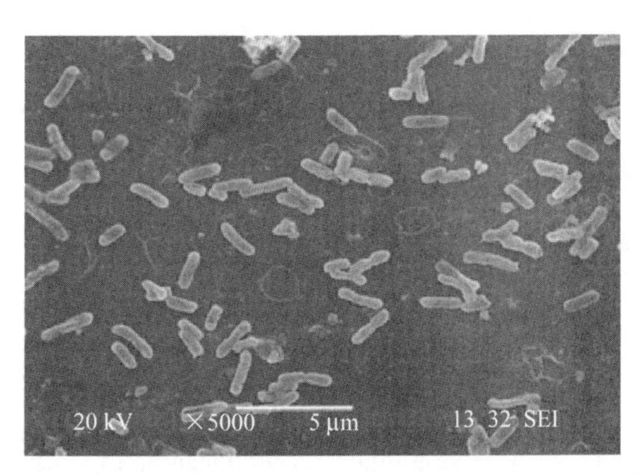


**Fig.2 The bioleaching microbes observed by SEM**

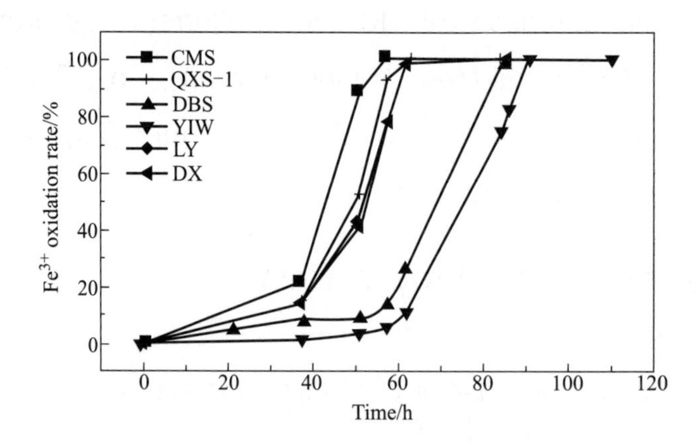


**Fig.3 The different bioleaching rate by several *A. ferrooxidans* strains with different oxidation and tolerance ability**

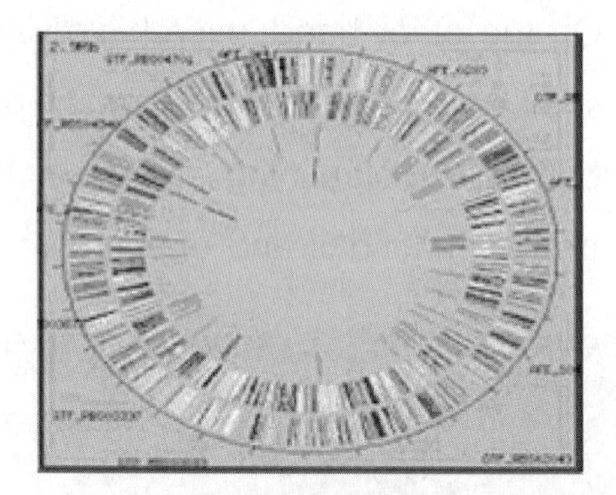

**Fig.4 *A. ferrooxidans* ATCC 23270 whole gene map**

In 2004, Central South University participated in whole genome sequencing of *Acidithiobacillus ferrooxidans* ATCC23270, which is the world's first sequencing for bioleaching microorganisms. Based on all of 3217 gene information obtained from the whole genome sequence (Fig.4), the whole genomic array and comparative genomics research were carried out. We found 320 high oxidation genes, whose 135 are related to ferrous oxidation, sulfur oxidation and resistance (Fig. 5). Based on this, a national standard (GB/T 20929 - 2007) " Methods for the detection of *Acidithiobacillus ferrooxidans* and its oxidation activity by microarray" has been established.

The establishment of national standard permitted to perform a rapid and accurate screening of bioleaching strains with high oxidation ability (Fig.6 and Fig.7). The whole genome map and annotations of *Acidithiobacillus ferrooxidans* laid the foundation of studying the bioleaching mechanism at gene level and realizing the orientation of microbial leaching behavior research from phenotypic to genotypic level.

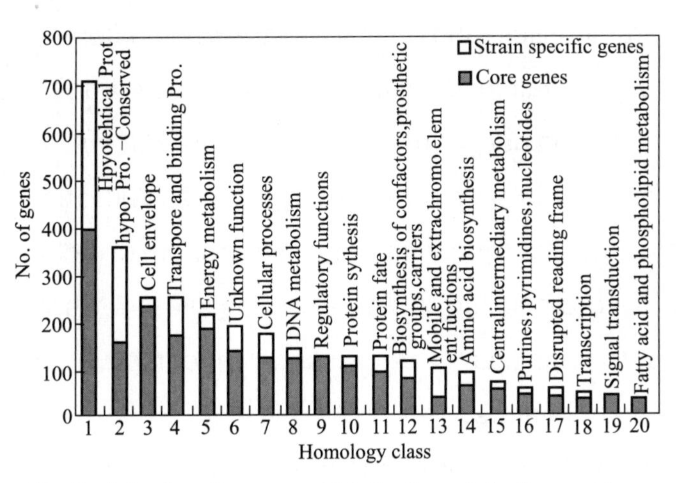


**Fig.5 The function gene distribution of *A. ferrooxidans***

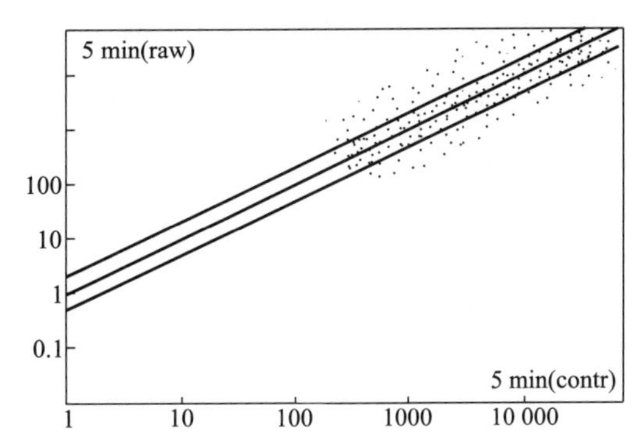


**Fig.6 Gene chip test of bioleaching strains(3 days)**

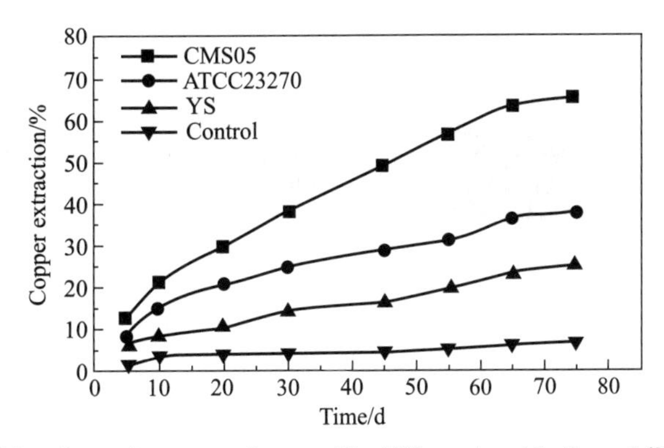


**Fig.7 Bioleaching by microorganisms with different oxidation ability (75 days)**

## 2.2 From the qualitative to quantitative analysis

It is the key to find a way to quantitatively analyse the microbial composition and function for the clarification of multi-factor strong correlation mechanism in bioleaching system, as the determination methods of bioleaching process conditions and chemical and physical parameters have been established. However, bioleaching microorganisms are various, their characteristics and functions are very different, their interactions are complex, their artificial cultivation is difficult, so that their traditional analysis methods based on isolation and biological chemical reactions are difficult for real-time quantitative monitoring and analysis of bioleaching microorganisms. With the rapid development of biological technology: genetic, genomic, and metagenomic technologies are more and more applied in biohydrometallurgy field; especially the applications of genomic technology have led to significant progress on quantitative analysis of bioleaching system with fundamental changes.

The development of bioleaching microbial function gene array and community genomic array technology, has led the research level from single function of single population to whole functions of single population and whole functions of a microbial community (Fig.8). Based on these technologies, the dynamics of microbial community structures and leaching functions can be detected quantitatively, which may be used to analyse the effect of leaching parameters on microbes growth and oxidation ability.

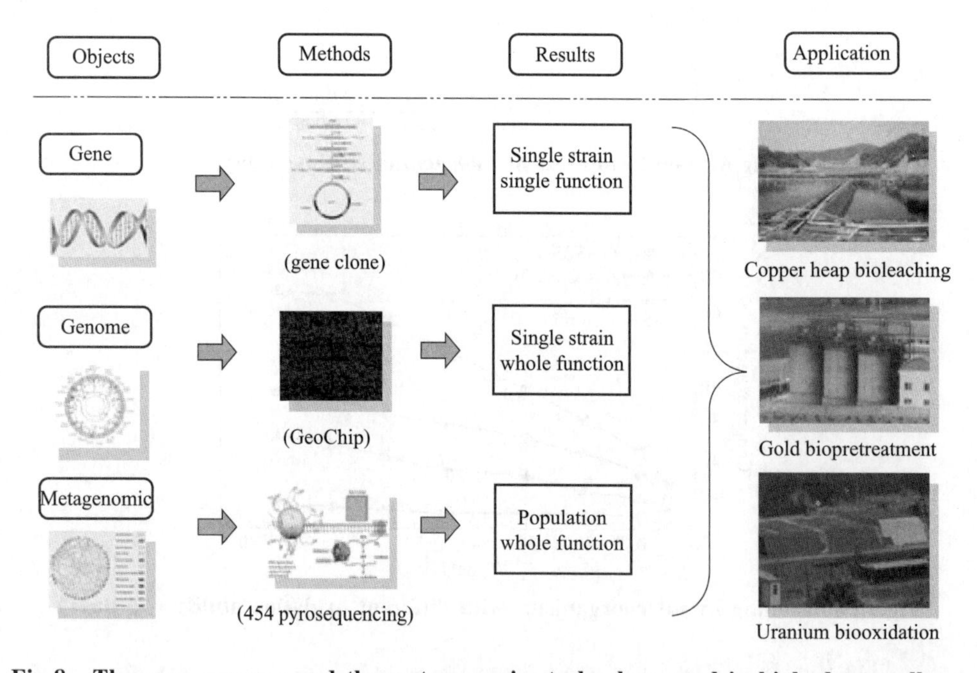


**Fig.8 The gene, genome, and the metagenomics technology used in biohydrometallurgy**

The established microbial function gene array was used to study the bioleaching microbes structure and function (Fig. 9 – Fig. 12), which solved the problem for synchronous detection of the microbial community structure and function in bioleaching system.

Based on the clarification of the succession mechanism of bioleaching microbial

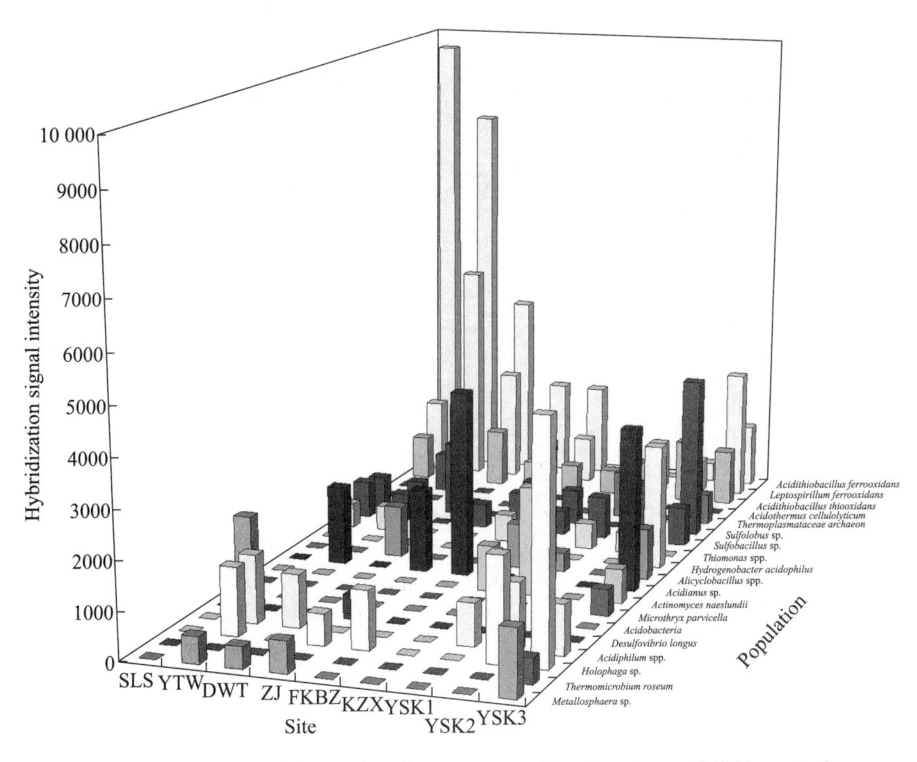


**Fig.9 The bioleaching microbes community structure of different site**

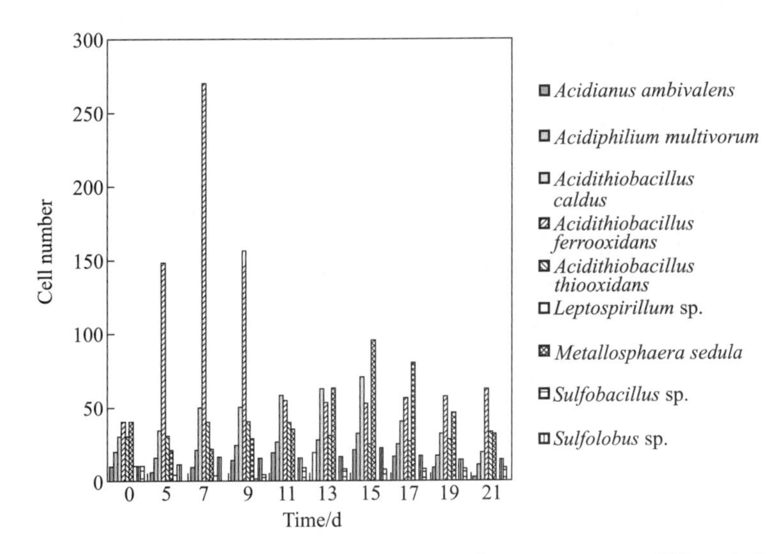


**Fig.10 The bioleaching microbes community structure at different time**

community structure and function, the microbial consortium was optimized through the mix of different types of microbes. The new optimized consortium was used for the bioleaching of low grade copper sulphide in YuShui Copper Mine, Guangdong Province, China, and the copper extraction was remarkably raised (Table 1).

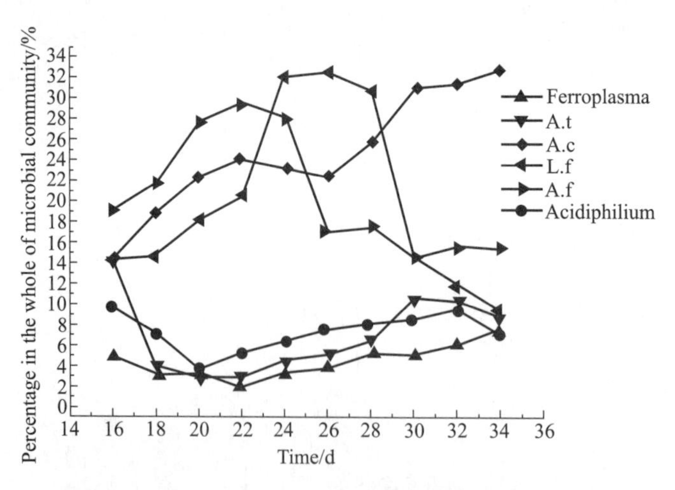


**Fig.11 The bioleaching microbial community dynamics at different time**

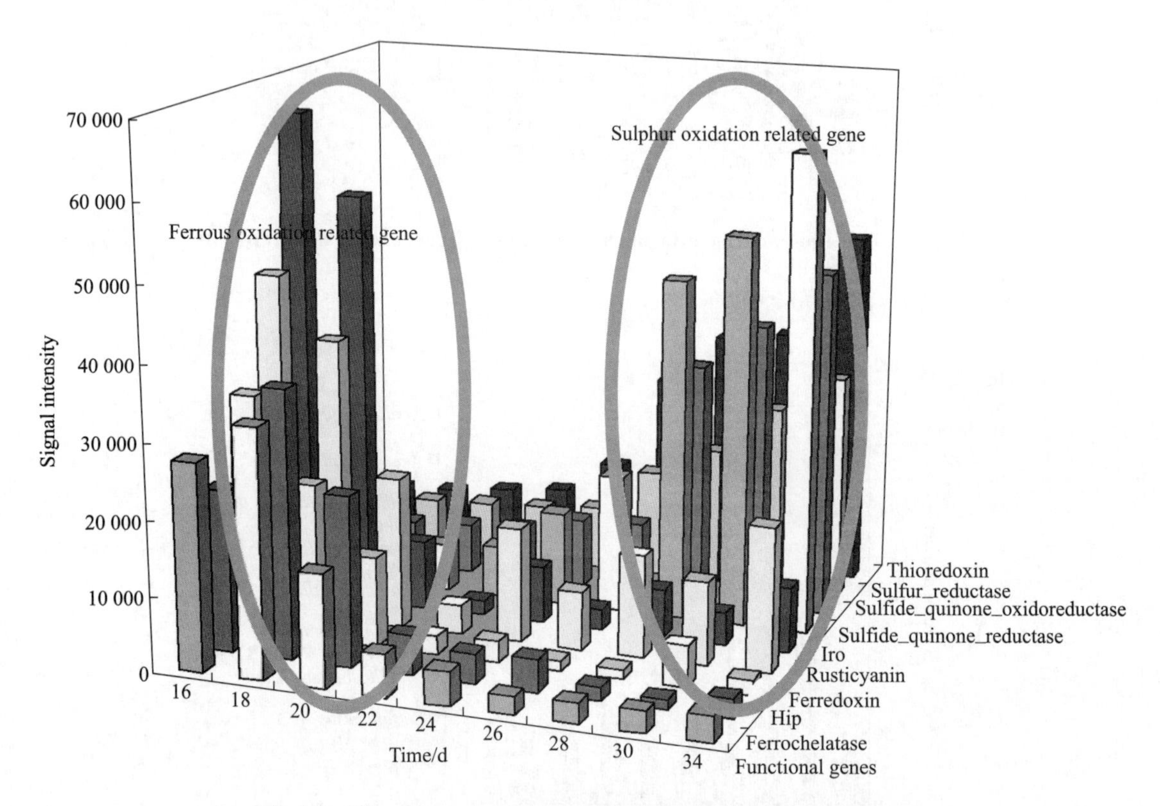


**Fig.12 The dynamic of bioleaching microbes function gene**

**Table 1 Four optimized microbial consortia and their demonstration effect in bioleaching application**

| Microbes | *A. f* | *Lep.* | *Acidip.* | *Actino.* | *Acido.* | Extraction | Time |
|---|---|---|---|---|---|---|---|
| Combination I | 17.5% | 23% | 8.7% | 1.6% | 10.3% | 55.23% | |
| Combination II | 37.5% | 22.7% | 4.5% | 9.8% | 7. 1% | 57.21% | |
| Combination III | 16% | 29% | 2% | 0% | 43% | 62.25% | 55 days |
| Combination IV | 12.1% | 16.9% | 38.7% | 1.7% | 5.8% | 75.11% | |

### 2. 3 From theory to practice

Based on the above researches, we successfully applied the results in industry for the extraction of copper, gold, uranium resources.

2. 3. 1 The bioleaching industry application in copper resource

1) Dexing copper mine in Jiangxi Province

The heap bioleaching plant in Dexing Copper Mine is an example of successful application of biohydrometallurgy technology. More than 350 million tons of waste ore have been released during the long term of mining process, and those wastes contain 0.05%–0.25% copper, so there are totally about 600 000 tons of copper metal. Since the waste ore are mainly composed of primary copper sulfide, it becomes difficult to obtain high leaching rate by using the traditional biohydrometallurgy method.

In order to effectively recycle Dexing copper waste ore, two projects about " Studies on bioleaching of low grade sulfide ore with artificial bacteria and the third phase extraction " and " Studies on the catalytic mechanism and strengthening bioleaching strains isolated from Dexing copper mine, and their industrial application" were carried out. Using the quantitive analysis technology of Central South University, bioleaching strains with high growth, high oxidation ability and high resistance to metal ions were obtained by microarray screening method. Furthermore, the SX–EW plant with 2000 tons of copper production was improved, and the copper extraction percentage and rate increased in a large scale (Fig.13).

**Fig.13 Biohydrometallurgy industrial application in Dexing copper mine in Jiangxi Province**

2) Zijin copper mine in Fujian Province

Zijinshan copper mine is a classic representative for the successful industrial application of biohydrometallurgy in China. Zijinshan Copper Mine is located in Shanghang County, Fujian Province, and detains proven reserves of 240 million tons of copper sulfide with 0.063% copper in average. Secondary sulfide ore is the priority mineral, and mainly contains chalcocite, covellite and enargite. Since the copper grade is very low, the traditional flotation and smelting process can't economically and effectively deal with these ores.

However, biohydrometallurgy is an alternative and feasible technology to effectively recycle this kind of resource. Through the experimental study of heap leaching industrialization (Fig. 14), Zijinshan Copper Mine has established a heap bioleaching factory with an annual capacity of 10 000 tons cathode copper. The available copper ore grade dropped to 0.40%, copper reserves increased from 2.7841 to 3.0746 million tons, the copper extraction percentage achieved up to 80.11%, heap leaching cycle decreased to 185 days, and the cost of high purity cathode copper products decreased to 12 812.00 Yuan/t Cu. The economic and social benefits of the enterprise are remarkable compared to traditional flotation and smelting process (Table 2).

**Fig.14 Heap bioleaching industrial application in Zijinshan copper mine**

3) Chambishi copper mine in Zambia

Biohydrometallurgy achievements developed by China were not only successfully applied in China, but were also spread into some foreign countries. In 2010, a strategic framework agreement was signed between Zambia Ministry of Mines and Minerals Development and Central South University, and " China Nonferrous Metal Mining (Group) Co., Ltd—Central South University Zambia biohydrometallurgy technology industrialization demonstration base" was established, which will promote the application of biohydrometallurgy technology to the treatment of lots of tailings and waste ore in Zambia.

**Table 2 The comparison of bioleaching and conventional smelting technology/ton. copper**

| Treatment methods | Resource/t | Energy consumption | | Water/$m^3$ | Green house effect | Acidification effect |
|---|---|---|---|---|---|---|
| | | Electric/(kW · h) | Coal/kg | | $CO_2$/kg | $SO_2$/kg |
| Biohydrometallurgy | 307. 46 | 3915. 14 | 1402. 22 | 21. 76 | 4090. 57 | 11. 93 |
| Flotation-smelting | 278. 41 | 8706. 90 | 3656. 24 | 168. 09 | 10 909. 29 | 79. 04 |
| Comparison | 110.43% | 44. 97% | 38. 35% | 12.85% | 37. 50% | 15. 09% |

In March 2011, Zambia Chambishi Copper Company cooperated with Central South University to exploit the low grade copper ore in Chambishi by heap bioleaching. Firstly, the local microorganisms were screened, enriched and domesticated for their adaptation to the heap environment, and kept good growth and oxidation ability. Then, the adapted microorganisms were expanded by culturing in 5 L, 50 L, 1 $m^3$, 20 $m^3$ and 150 $m^3$ stirred tank, successively (Fig. 15). The microbial consortium were inoculated into ore heap, and the cell concentration significantly increased in the heap leaching system, and the copper extraction rate was improved and the leaching time was shortened (Table 3).

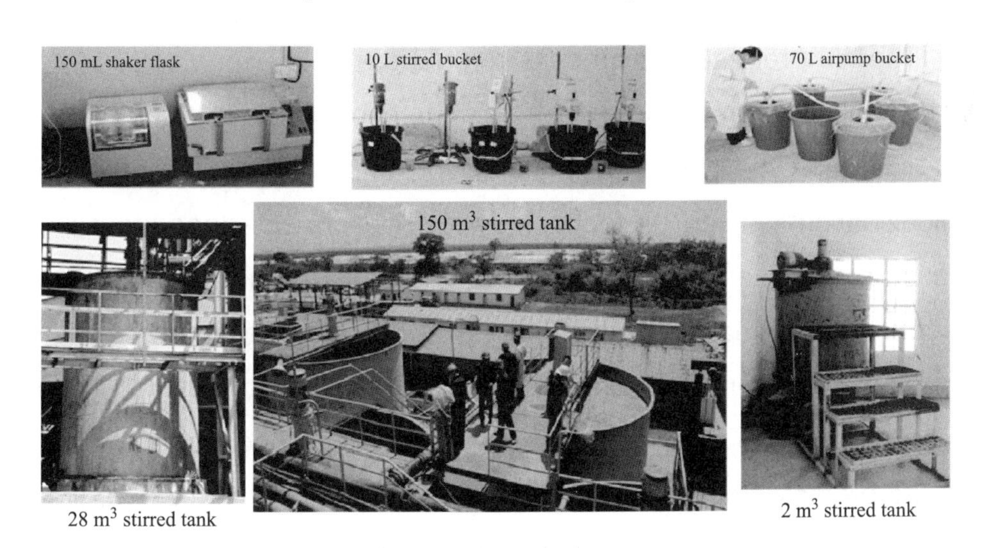


**Fig.15 The expanding culture process of bioleaching microorganisms in Chambishi copper mine in Zambia**

**Table 3 The comparison of the existing biohydrometallurgy and the original hydrometallurgy process**

| Parameters | Hydrometallurgy | Biohydrometallurgy |
|---|---|---|
| Ore grade / % | ≥1.2 | ≥0.3 |
| Leaching time | 4-8 h | 60 d |
| $H_2SO_4$/ (g/L) | 30-40 | 10-15 |
| Microorganisms | — | Optimized combination |
| $Cu^{2+}$/(g/L) | 4-8 | 3-5 |
| Extraction percentage /% | 65-70 | 85-90 |
| $H_2SO_4$/t copper | 4.8 | 2.2 |
| pH | 1.4-1.6 | 1.8-2.1 |

In the heap bioleaching of 600 000 tons of low grade copper ore, copper extraction percentage could achieve up to 50% in 2 months. Bioleaching solution was treated by the subsequent SX-EW process, and 99. 99% purity of cathode copper was obtained. Without changing the main original engineering conditions, the existing biohydrometallurgy technology could make the heap-leaching copper production increased by 20%, and the acid consumption reduced by 35% or more (Table 3), compared to the original hydrometallurgy process. Hence, by using biohydrometallurgy technology, a large number of low grade copper resources could be effectively recovered.

2. 3. 2 The bio-pretreatment process of gold ore

China National Gold Group Corp. works on the exploitation of arsenic refractory gold concentrates in Tianli Gold Company, Liaoning province, China. They did lots of works based on biooxidation-cyanidation technology, carried out self-dependent innovation on the biooxidation process, and finally developed the CCGRI gold biooxidation technology with independent intellectual property rights. The biooxidation microorganisms (called HY series-bacteria) used in this technology were cultured and adapted for a long time and are very different to the microorganisms in BIOX and BACOX process. This technology was promoted and exploited, and the Chinese gold industry's first new high-tech industrialization demonstration project was constructed in 2003. Since then, the operation was stable, and has been working about 8 years. The production and technical index have become more and more advanced. With the development of the gold ore biopretreatment technology, China's gold production has become the world's first for four years (Fig.16).

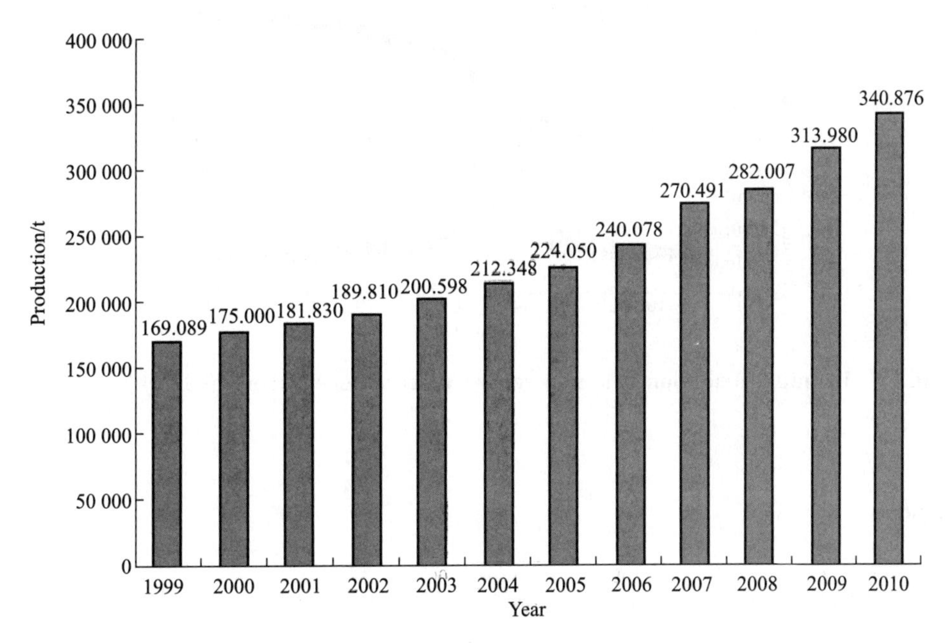


**Fig.16 China's gold production has become the world's first for four years, with the development of the gold ore biopretreatment technology**

2.3.3 Bioleaching process of uranium ore

In order to be in pace with the increasing demand of uranium that the development of nuclear power needs, China's uranium production has been oriented to the exploitation of low grade or refractory uranium ore, and other mineral resources associated with the processing uranium. During acid leaching of uranium ore, the tervalent uranium ion must be oxidized to pentavalent uranium ion by ferric iron, leading to the consumption of ferric iron. Hence, the bioleaching technology used in the treatment of uranium ore allows the ferric iron regeneration, and therefore keeps the leaching reaction continuous.

Bioleaching technology will play an important role in uranium ore processing. Using the optimized mixed culture into the heap leaching in Fuzhou 721 mine in Jiangxi, the uranium extraction percentage could achieve up to 96.82% (Fig.17) in 97 days. The promotion and application of biohydrometallurgy technology can make large number of idle or abandoned uranium sulfide resources available in China. It is expected to make uranium resources exploitation grade decrease from the current one over one thousand to three over ten thousand, and thus the economic mining reserves of uranium ore can be greatly increased.

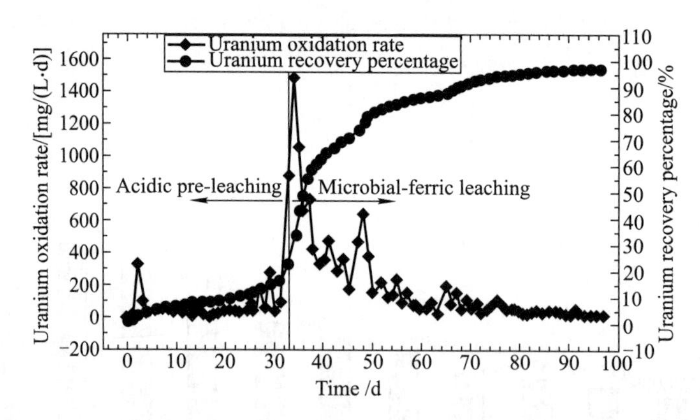


**Fig.17 Uranium oxidation rate and recovery percentage during bioleaching process**

## 3 Outlook

The conventional metallurgy process involving intense oxidation and reduction reactions is a process which requires a lot of energy consumption and produces a large amount of carbon, and then causes serious environmental pollution. While biohydrometallurgy reactions mainly occur under normal temperature and pressure,and involves mild oxidation and reduction reactions. Furthermore, during bioleaching, the microorganisms use carbon dioxide from the air to synthesize organic substances,and this process does not produce carbon dioxide to pollute environment,instead,as plants it consumes carbon dioxide in the atmosphere,and thus is a carbon reduction process.

Biohydrometallurgy technology is a clean, safe, and low cost way: a. It can effectively expand the available resources: raw ore,waste ore,tailings,slag,combustion ash,secondary metal,industrial electronic junk,sewage sludge,etc. As long as there is valence changes,bioleaching technology can be used for the leaching of copper,gold and uranium ores,including sulfide ore,oxide minerals and complex mineral;b. It can be widely applied ① in static leaching (such as heap leaching) for copper,gold,uranium and nickel extraction,and ② in dynamic leaching (such as stirred bioleaching) for gold, copper,nickel and cobalt extraction.

Therefore,it is imperative to find a new mineral processing way to deal with these low grade resources economically and environmentally friendly, and in this case, biological technology stands as an efficient alternative,and can be the key to unlock mineral resources value.

**Prof. Guanzhou Qiu** born on February 2, 1949 in Guangdong Province, China; graduated from Central South University of Technology in 1987 (now called as Central South University) and received his Ph. D. in Minerals Processing; was selected as the Academician of Chinese Academy of Engineering in 2011.

Lecturer, associate professor and professor of Central-South University (CSU); vice director, director of Department of Minerals Processing (CSU) (1987—1992); Vice President of CSU (1992—2010); Chairman of the 19th International Biohydrometallurgy Symposium (2010); Vice President of International Biohydrometallurgy Society.

Major Subjects: mineral flotation and flotation reagent, electrochemistry of flotation for sulfide mineral, flotation of fine particles, agglomeration and dispersion of fine particles, solution chemistry of flotation, biohydrometallurgy for sulfide mineral, applied surface chemistry in mineral, metallurgical and material processing, waste and secondary resources recovery etc.

He gained 4 national prizes, and over 10 national key projects; was authorized for 42 patents, and published 5 books and over 100 papers as the first or corresponding author.

# Systems Approach for Process Excellence

**Arthur Ruf**

ETH Zurich, Zurich, Switzerland

**Abstract**: The huge amount of information on specific topics in different fields-chemical, metallurgical and material engineering-is very impressive and an excellent basis for the improvement of existing processes and also for innovation. The most important issue in the future is the ability to create knowledge out of this vast field of information. Knowledge which can add to the solution of industrial questions in the search for competitive advantages. The deep understanding in fluid dynamics, heat and mass transfer, kinetics, sensor technology, IT, modelling and simulation and material science is the precondition for chemical, metallurgical and material engineering. On the basis of industrial cases the future key elements for the chemical, metallurgical and material engineering is derived.

Dr. Arthur Ruf Vice President & Head of foreign relations, Swiss Academy of Engineering Sciences (SATW), ETH Zurich.

1981 Dr. sc. techn. ETH

1981-1989 oberassistent at the Institute of Process Engineering / ETH

1984-1996 lecturer at the Dept. III/A of the ETH, lectures on "Membrane Separation Technology"

2004 lecturer at the University for Applied Science Chur Conferences

1995 chairman of the Organizational Committee of R'95" (February 1995 in Geneva with more than 1000 attendees)

1997 member of the Steering Committee of R'97,99,01,05,07,09

2003 chairman of the ACHEMA in Frankfurt/Main

2005 chairman of the "Industrial Committee of R'05" in Beijing

2007 chairman of the "Industrial Committee of R'07" in Davos

2009 chairman of the "Industrial Committee of R'09 and WRF" in Davos Industry

1989 start at Bühler AG in 9240 Uzwil (Switzerland)

1989–1996 head of R&D of a BU, division (60 personnel)

1996–2002 head of division, member of the Group Management (800 personnel) Organizations

–Treasurer of the ESMST (European Society of Membrane Science and Technology)

–Member of the board and responsible for "Foreign affairs" of the SATW (Swiss Academy of Technical Sciences)

–Personal member (fellow) of the "SATW" (Swiss Academy of Technical Sciences)

–Member of the "DECHEMA" (Germany)

–Vice president of the SATW

Independent company

2002 Founder of the engineering / consulting company "4p & partners"

–Member of the board in different companies

–Mandates in business development

–Management on time

–Support in strategy processes and business processes

–Technology support

–Partner InnoVAVE